MERSEYSIDE and WEST C[ ]
RAILWAYS 1965-1990

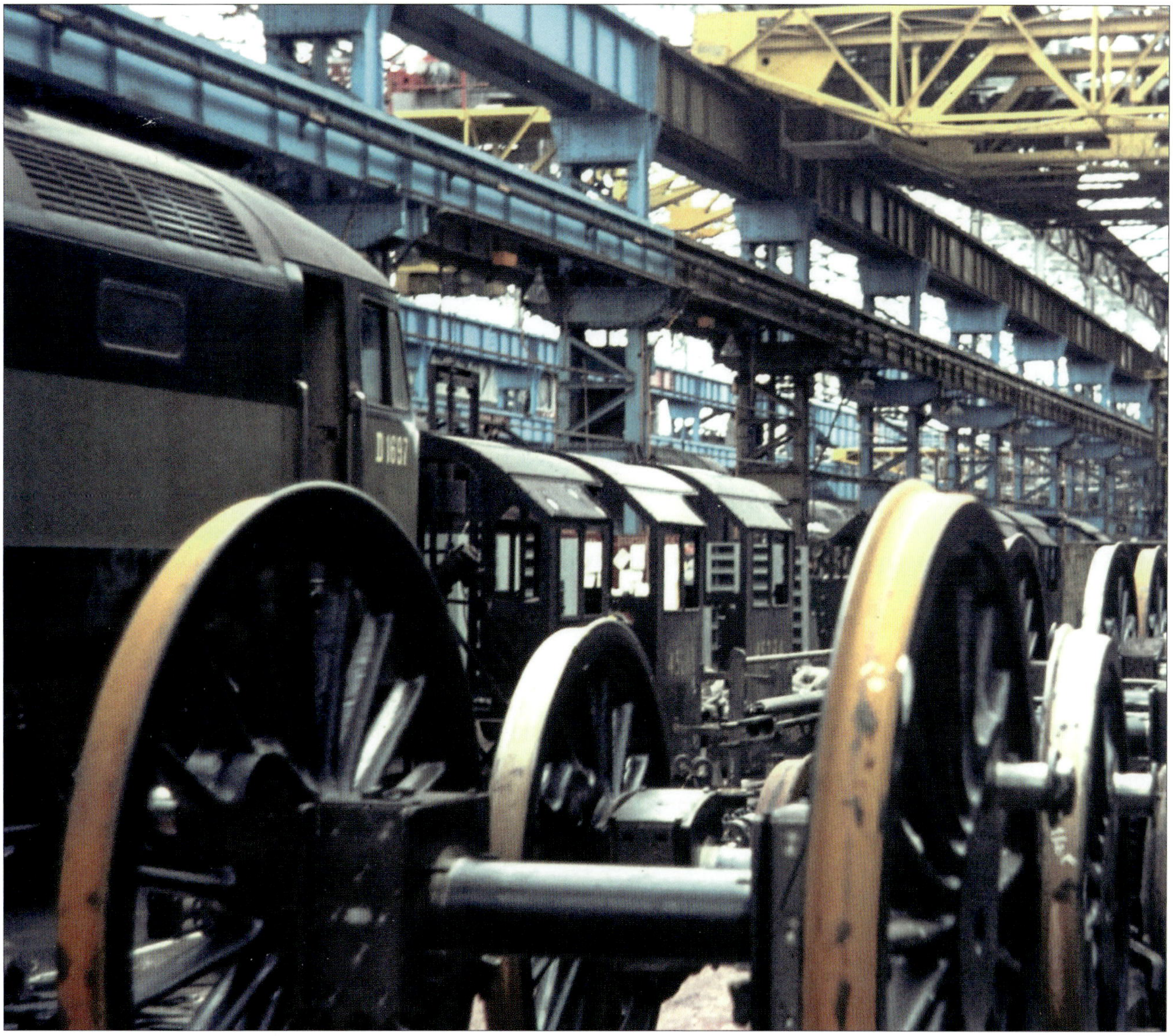

Henry Finch

Merseyside and West Cheshire Railways 1965-1990

Reviving the memories of yesterday…

© Design: The Transport Treasury 2024. Images and text: Henry Finch

ISBN 978-1-913251-70-3

First published in 2024 by Transport Treasury Publishing Ltd., 16 Highworth Close, High Wycombe, HP13 7PJ

www.ttpublishing.co.uk

Printed by Short Run Press, Exeter, UK

The copyright holders hereby give notice that all rights to this work are reserved. Aside from brief passages for the purpose of review, no part of this work may be reproduced, copied by electronic or other means, or otherwise stored in any information storage and retrieval system without written permission from the Publisher. This includes the illustrations herein which shall remain the copyright of the copyright holder.

Front cover top. A Stanier 8F 2-8-0 crosses the River Weaver at Frodsham in August 1966 (see also p.74)

Front cover lower left. Steam met diesel in May 1966 near Hoole Lane, Chester. 'Black 5' 4-6-0 no. 45373 was arriving from Crewe, with EE diesel no. D255 heading for Crewe from North Wales.

Front cover lower right. A DMU leaves Olive Mount cutting with a local service from St Helens to Lime Street in July 1983 (see also p.30)

Back cover. At the southern end of Moore troughs, viewed from the footbridge over the WCML on 17th June 1967, Standard 9F 2-10-0 no. 92051 is passing with a train of vans. The amount of coal in the tender is looking rather low so the 9F will probably come off at Crewe, but the driver decided to make use of the troughs to ensure enough water and they've overdone it. The drama of such a cascade might have been its impact on the front vehicle of a passenger train running at express speed with ventilators open on a hot summer's day, but this overflow is no worse than untidy. On the distant skyline directly above the loco's smoke deflector is the Keckwick bridge which carries the Warrington – Chester line over the WCML.

Frontispiece. The erecting shop at Crewe Works on 27 February 1966 held a variety of half-stripped locos and a clutter of bits and pieces strewn all over the floor. On the left in two-tone green livery is Brush D1697, a class 47 Co-Co diesel electric later renumbered 47109, built in 1963 and withdrawn in 1987. It was sharing space with three cabs from Stanier locos, the furthest belonging to 'Black 5' 4-6-0 no. 45234. The foreground is dominated by three sizes of wheelsets, and nearby lay a confusion of connecting rods and smokebox doors. Two 9F 2-10-0 locos were also present receiving attention, no. 92208, and one of the former Crosti-boilered group. The 'Crosti' reminds me that about seventy years ago a cousin and I were treated to a mid-week excursion from Watford Jct to Crewe Works with a conducted tour. The din, the fumes, and the physical activity on display were a lot to take in. But I vividly recall the sight of molten metal being poured into moulds in the foundry, and a pile of castings with chalked numbers on them beginning 9202... The first 32 Standard 9Fs had been built before the 'Crostis' (nos. 92020-9) appeared in March-July 1955. Were the castings in fact of new components found necessary for the 'Crostis' once in service?

Contents

1 LIVERPOOL	5
2 WIRRAL	41
3 CHESTER	61
4 WEST CHESHIRE	70
5 West Coast Main Line (WCML)	91
Abbreviations	112
Sources and Bibliography	112

Introduction

Towards the end of 1963 I set foot in Liverpool for the first time, having arrived cross-country at Manchester Piccadilly and been advised to take a taxi to Manchester Central to continue by rail to Liverpool. It was an unfortunate interruption to my journey between the two great cities linked by rail since 1830. About to start my first paid employment I had enough other matters to concern me, but soon enough as a locospotter since the age of nine I started to take a lively interest in my new surroundings. The first step was to save enough to buy a camera. Finding a top-floor flat in Botanic Road for a year gave me views of Edge Hill and its Gridiron. Through the pages of *Trains Illustrated* I had admired the work of the masters, and in August 1966 I stood shoulder to shoulder on the bridge in York station with a Leica-bearer and ex-army knapsack who suggested after the Deltics passed that we get a cup of tea. I'd guessed already, but after twenty minutes not talking about railways, cameras or the Church he introduced himself as Eric Treacy. Truly a brief but treasured encounter.

This selection of my photographs is intended to record ordinary locomotives and trains performing everyday tasks, so I've drawn the line at enthusiasts' specials and preserved locos. An A4 on the WCML, and even a gleaming Castle on the Wirral, have been left in the locker. The classic three-quarters front view of a train can be uninspiring unless the circumstance or context of a scene offers something extra and that was always the aim. The existing literature available to latecomers like me was an enormous aid to making the selection, and my gratitude to all its authors is recorded in the Bibliography. In particular Joe Brown's magisterial *Liverpool and Manchester Railway Atlas* was a constant companion and source of enlightenment. Needless to say, all the errors are mine.

The photos are organised into five geographical groups though some overlaps are inevitable and I've taken a few liberties: Liverpool, Wirral, Chester, West Cheshire (to include Widnes, Runcorn and Northwich), and West Coast Main Line (WCML). Where possible I've tried to give some historical background and an indication of what may have changed since the photo was taken. Ideally there would have been more equal coverage to the two and a half decades, but there were periods when photography had to take a back seat because of work and family responsibilities. For those who find a mix of monochrome and colour photos confusing, it was a question in those pre-digital days of whichever film was in the camera at the time rather than deliberate selection. Final point: I abandoned railway photography in 1990 basically because of impending privatisation. BR was a major institution centrally administered for the nation as a whole when I first began to appreciate and enjoy observing its operation. Thinning out over time was inevitable, but BR stayed essentially an integrated network of routes and lines to all parts of the country. That is how the child in me would have had it remain.

Merseyside and West Cheshire Railways 1965-1990

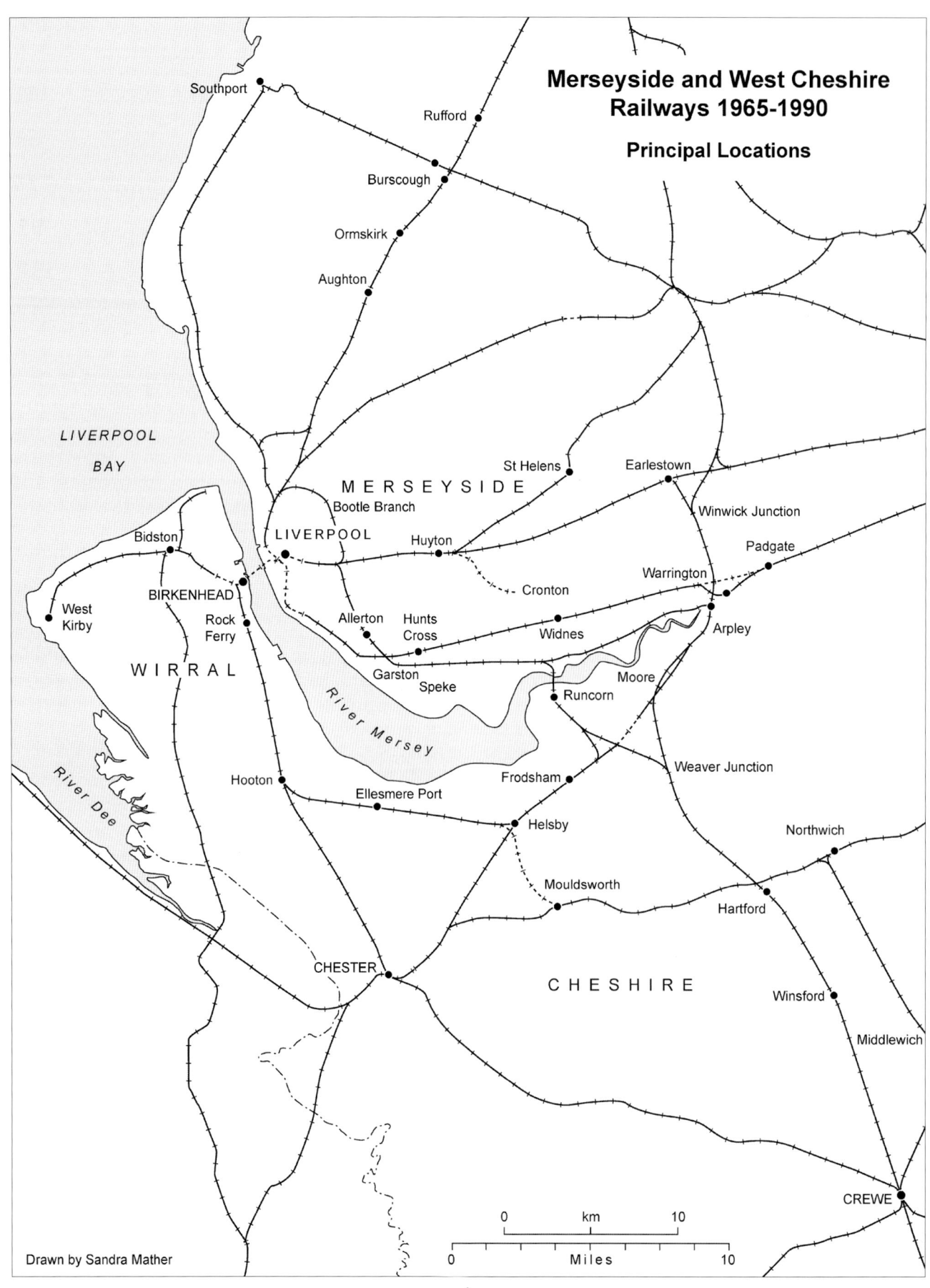

Merseyside and West Cheshire Railways 1965-1990

1 LIVERPOOL

Although Exchange Station, formerly the Lancashire & Yorkshire Railways' Liverpool terminus, was looking down at heel in the 1960s it did not close completely until 1977. That year Merseyrail opened Moorfields underground station (close to Exchange) on a loop line connecting to Liverpool's only surviving main line station, Lime Street. At the same time Exchange's pre-1914 electrified lines to Southport and Ormskirk were buried in a new tunnel through Moorfields which directly linked these towns north of Liverpool to Garston in the south, to be known as 'Northern Line'. On 1st July 1966 Jubilee class 4-6-0 no. 45627 (formerly *Sierra Leone*) brought in the empty stock to platform 3 for the 13.27 departure for Glasgow. Behind the Jubilee is a first generation diesel multiple unit (DMU), perhaps a class 108, on a local service. The diagonal yellow stripe on the Jubilee's cab side was a reminder that the loco was not permitted to work south of Crewe after August 1964 because of inadequate clearance under the 25kV electrification. Coal on its tender was piled high so there was work to be done later, but a Stanier 'Black 5' 4-6-0 no. 44809 left on time with the 13.27. No. 44809 was formerly a Southport (8M) loco but the shed had closed in May with its allocation presumably transferring to Bank Hall (8K), which (like loco no. 45627), struggled on until October 1966. *HF-S710*

Merseyside and West Cheshire Railways 1965-1990

Top left. By 1859 the L&Y had absorbed the East Lancashire and Liverpool Crosby & Southport Railways, and the original Exchange station of the 1850s in Tithebarn Street was extensively rebuilt in the 1880s. It possessed a roof comprising four linked glazed ridge structures covering ten platforms, each platform bay having two platform tracks except for road access between 3 and 4, and a third track between 6 and 7. During the Second World War the strategic port of Liverpool and its associated network of railways were obvious targets for the Luftwaffe, and Exchange Station was severely damaged in 1941. The two glazed roof sections covering platforms 6-10 were hit by bombs and they remained in their reduced state until final demolition in the 1980s. Only Exchange's splendid façade looking out on Tithebarn Street has survived, while the site still awaits redevelopment. On 26th March 1966, 'Black 5' 4-6-0 no. 44893 was making an impressive departure from platform 5 with the 17.40 service probably to Rochdale. Alongside in the picture on platform 4 is a DMU while to the right on the electrified lines is one of the 1938 LMS class 502 electric trains. *HF-S276*

Bottom left. Between platforms 3 and 4 a short section of track served as a bay that was sufficient to accommodate Exchange's station pilot. On the 1st March 1966 this was no. 41211, one of three of Bank Hall's Ivatt class 2 2-6-2T locos. Exchange in earlier times had two turntables, one located to the west of all the running lines, the other directly in front of the bay with only a short stub alongside it so giving no direct access for an arrival loco to turn. To the left of this turntable is a flat-roofed four-square brick office which partly hides a fine brick and timber signal box which controls the station area. This carries the name Liverpool Ex No2 where one might expect No 1, but it was built as Liverpool Exchange A box in 1886 and became No 2 in 1946 with partial adoption of colour-light signalling, when Exchange B (located where the four approaching lines became eight for entry to the station) was renamed No 1. It is remarkable (as it appears) that No 2 box could have survived the Blitz to the extent of merely needing repair rather than replacement. *HF-2-24-22-3*

Above. On 25th August 1967 'Black Five' 4-6-0 no. 44864 drew into Exchange station's platform 7 with the empty coaching stock (ECS) for the 13.27 Glasgow. The view shows the slender, elegant columns and complex ironwork needed to hold the roof section in place, but also gives a sense of Exchange station's decline. There are three slots for train information for the benefit of passengers, but only one is in use (though the same facility for platforms 8-10 by the loco's chimney would surely have been busier). Fortunately the clock seems to be in good order. The two ticket-collectors' cabins above the buffers would have been sought-after if salvaged when the station closed ten years later. Earlier in 1967 platforms 1 to 3 were taken out of use and converted to car parking, hence the platform change to 7 for the Glasgow train. Platforms 6 and 7 were the only pair to be separated by a siding, used primarily to store coaching stock at off-peak times. By 1967 the siding's usefulness was greatly diminished, and on this occasion it was occupied by Standard Britannia class 7MT 4-6-2 no. 70023 (formerly *Venus*), waiting for the arrival of the ECS at platform 7. *HF-5-58-23*

Above. On 5 May 1966, in only their second season in European competition, Liverpool FC played their first European cup final, at Glasgow's Hampden Park that evening. Their opponents in the Cup Winners Cup were Borussia Dortmund of West Germany, who won 2-1. Five football excursions to Glasgow were arranged from Exchange station that morning. The first was seen to depart at 10.30 having Stanier 'Black 5' 4-6-0 no. 45412 in charge of ten well-loaded coaches, a significant burden for a class 5 loco. The second at 11.10 was hauled by one of Bank Hall shed's last two Jubilee class 4-6-0s, no. 45627. The third 'footex', seen here with black exhaust and a plume of steam against a threatening grey sky, was another of the numerous 'Black 5s' (possibly no. 45065 though with the final digit obscured by its reporting number 1X30). The two other specials were also hauled by 'Black 5s'. The train was about to pass Sandhills, the first station (about 1½ miles) out from Exchange. Beyond Sandhills, the line to Southport (just visible on the far right of the picture) begins to diverge from the main line to Preston and the WCML, but all lines at this point are electrified hence the urgency of the warning signs. *HF-S404*

Top right. The Britannia class 7MT was the first of BR's Standard classes to appear in 1951, with 55 of the type having emerged by 1954. They were distributed widely to BR's regions, with the Great Eastern section of the Eastern Region (limited to class 5 power at the time) first in line for twenty of the new arrivals. The Western Region's share was fifteen, of which no. 70023 was initially allocated to Old Oak Common. Like all the other Britannias, no. 70023 now stripped of its nameplate ended its days on the London Midland Region (BR's last regional steam user) and on 25th August 1967 was one of only 20 still active at the time. It was photographed from Stanley Road bridge just north of Sandhills, well in command of its light-weight 5-coach load forming the 13.27 Liverpool-Glasgow service. The burden would be much greater when joined by the Manchester section of the train. In the foreground is Kirkdale South Jct, where the lines to Bootle and beyond to Southport diverge to the west, while the main line continued towards Ormskirk and Preston with the Wigan and Bolton routes diverging eastwards at Walton Jct. Notice how well the Bankhall area (and, further on, Chapel Street Southport) is supplied with bolts and nuts. *HF-S1990*

Bottom right. Earlier, on 1st July 1966, Jubilee class 4-6-0 no. 45627 had drawn ECS into Exchange station, and is seen here returning to Bank Hall shed (8K, recoded from 27A in 1963) to await the next job. (With the withdrawal of long-term resident no. 45698 *Mars* in November 1965, no. 45627 was the shed's last class 6 loco.) The viewpoint is from the north side of Stanley Road, with the coaling tower prominent to the left with Sandhills No. 2 signal box in the centre. Between them is a small shed, probably a three-road repair shop at one time but visible in the photo are an oil tank, and a Stanier loco in store with its chimney wrapped, ready for the call that never came. The shed site, with the main shed hidden beyond the repair shop, lay in a triangle between the lines to Southport and to Kirkdale, with the third side of the triangle formed by the LNWR's Bootle Branch line of 1866 which gave the LNW access from Edge Hill to the docks, and later allowed through services from Southport via Lime Street to Euston. After the shed's closure, ten years later it was rebuilt and renamed Kirkdale Depot for servicing Merseyrail's EMU classes 507 and 508 and the long-awaited Stadler EMU class 777. *HF-3-58-14*

Merseyside and West Cheshire Railways 1965-1990

Merseyside and West Cheshire Railways 1965-1990

Top left. North from Kirkdale the line to Preston ran through Ormskirk and was electrified by the LYR that far in 1913. The Beeching Report in 1963 called for the closure of both the Exchange-Southport line and the third main route from Exchange via Kirkby to Wigan, but proposed to maintain the through route from Exchange to Preston. In the event the closures were substantially reversed: the electric lines to Southport, Ormskirk and (since 1977) Kirkby have all been integrated into Merseyrail's Northern Line. That has left doubt about the future of lines beyond the electrified sections, most notably in Ormskirk where the line has been severed with back-to-back buffers, and the line on to Preston singled. In happier times the 11.30 Glasgow - Exchange is seen having just passed Aughton Park station (south of Ormskirk on the third-rail electrified section) behind 'Black 5' 4-6-0 no. 44872 on 16th July 1967. The loco was newly-arrived at Lostock Hall (10D) from Aston (2J). It had previously had (and lost) its moment of glory, entrusted with the 08.15 Nottingham-Marylebone service on the final day of through working on the GC line, 3rd September 1966, but failed and had to be replaced at Aylesbury. Chalk markings on the smokebox door suggest it had since received another opportunity to work a special or extra train. *HF-S1868*

Bottom left. Burscough town had a population of just over 9,000 in 2011 but had two railway stations on two railways, a congested main road problem and a junction on the Leeds & Liverpool Canal. Burscough Junction is the first station from Ormskirk on the line to Preston, a half-mile from Burscough Bridge station on the Southport to Wigan line which passes underneath the former nearly at right angles. The lines were previously connected by the North and South Curves which are visible in the photo looking east from the A59 road bridge on 16th October 1965. A WD class Austerity 2-8-0 is heading north towards Preston with a trainload of permanent way (p.w.) sections. Riddles based the 1943 design on Stanier's class 8F 2-8-0 but opted for the parallel boiler and round-topped firebox to reduce construction costs. It's a good guess that the loco is from Aintree shed (which had 14 WDs on its books a year previously) and collected its burden from nearby Fazakerley p.w. depot. Next day was a Sunday so the new track would be needed for track engineers to take advantage of the reduced Sunday timetable. Both curves have now been lifted, but the South Curve is remembered as the site of a collision near Junction station in 1880 which claimed nine lives including three of the four footplatemen. *HF-1-60-31*

Above. Two stops north of Ormskirk is the small community of Rufford, where a branch of the Leeds and Liverpool Canal keeps close company with the railway. On 16th October 1965 the 17.12 service from Preston to Liverpool Exchange was drawing in to Rufford station past the signal box hidden behind class 4MT 2-6-4T no. 42187, one of about 250 of the type that were built between 1945 and 1950 for the LMS and BR. The design of this tank engine is generally attributed to Fairburn who succeeded Stanier as CME of the LMS in 1944 but died the following year. The unfortunate Fairburn made only small changes to Stanier's own design – mostly to the evaporating surface area and superheat, none of them very visual – and was in any case throughout his career more interested in the development of the diesel-electric locomotive. As the photo shows, Rufford station was picturesque but quite primitive in 1965, built of wood with gas-type lights. But at least it had a through service. Now the old station and signal box have been demolished and replaced by basic concrete structures and a crossing barrier in place of gates, and Rufford has been the only crossing point on the single-track line between Ormskirk and Preston. *HF-1-59-27*

Above. On the 14th August 1965 Southport station (from memory still trading as Chapel Street though the CLC's Lord Street had long since closed) welcomed the arrival of the 12.15 Manchester Victoria - Southport train hauled by 'Black 5' 4-6-0 no. 44686. This engine and its twin were especially significant among Southport (8M) shed's collection of Caprotti locos as the last 'Black 5s' be built, at Horwich in 1951. The success of British Caprotti valve gear on nos. 44686/7 led to its adoption on thirty of the BR Standard 5MT 4-6-0s (nos. 73125-54) in the later 1950s. The photo of no. 44686 backing out of the station gives a sleeves-rolled-up image with its double chimney, high running plate, and cab sides cut short (as with the last two Coronation 4-6-2s). Two months later no. 44686 was withdrawn from service. Two months earlier the Southport part of the route built by the Manchester and Southport Railway in 1855 running directly into Chapel Street station was closed from Pool Hey Jct, with traffic thereafter taking the loop round via Meols Cop into the station. On the day of the visit (a Saturday) BR was holding an auction there of railway memorabilia which included a double chimney probably from a dismantled withdrawn double-chimney 'Black 5'. Its considerable weight discouraged any casual bids. *HF-1-37-33*

Top right. Liverpool's first railway station, the terminus at the western end of the Liverpool and Manchester Railway, was built in 1830 at Crown Street, reached by a short tunnel from Edge Hill. (Also in 1830 a separate but adjacent tunnel of about 1¼ miles ran to Wapping to receive goods from the dock.) But Crown Street station ('The Railway Office, Liverpool' in T.T. Bury's 1831 lithograph print) was too distant from the city centre, and by 1832 work was underway for a new terminus at Lime Street. The railway site at Crown Street was left to operate as a coal distribution yard, and in May 1966 it was found with a 'Jinty' 3F 0-6-0T no. 47406 in attendance, with the informal title 'King of Edge Hill' chalked on the side. Coal is evidently being bagged up in a mineral wagon and manually loaded onto a William Fowler & Sons (a Liverpool coal merchant) flatbed lorry. The loco's reign was short-lived being withdrawn from service at the end of 1966, but after a prolonged stay at Barry scrapyard it was restored in working order. The Crown Street site was closed in 1972 and is now grassed over as an open space for public enjoyment, with little or nothing to indicate its historic importance. *HF-S426*

Bottom right. Lime Street station opened in 1836 as the new terminus of the LMR. It was approached from Edge Hill by a new 1¼ mile tunnel on a gradient of 1 in 93, with trains minus loco descending by gravity and hauled back up by a stationary steam engine. The single tunnel later became punctuated by open stretches of cuttings, rope was displaced by steam locos in 1870, and two tracks became four in 1881. The station itself urgently needed to expand, with its first train shed completed in 1851 by when the LMR had been absorbed by the Grand Junction Railway (GJR) and shortly after by the London and North Western (LNWR). That first shed was replaced in turn in 1869 by Lime Street's first arched train shed – on the station's northern side – followed in 1879 by a matching adjoining second shed on the southern. This second arch is visible from the street outside, but the face of the northern arch has been completely concealed since 1871 by the North Western Hotel. On 20th August 1966 'Black 5' 4-6-0 no. 45249 was under the northern arch waiting to leave with the 10.35 (summer Saturdays only, 1D26) train to Caernarfon, a service which ended altogether with the closure of the Caernarfon branch in 1970. To the right of the picture is a Trans-Pennine class 124 DMU. *HF-S920*

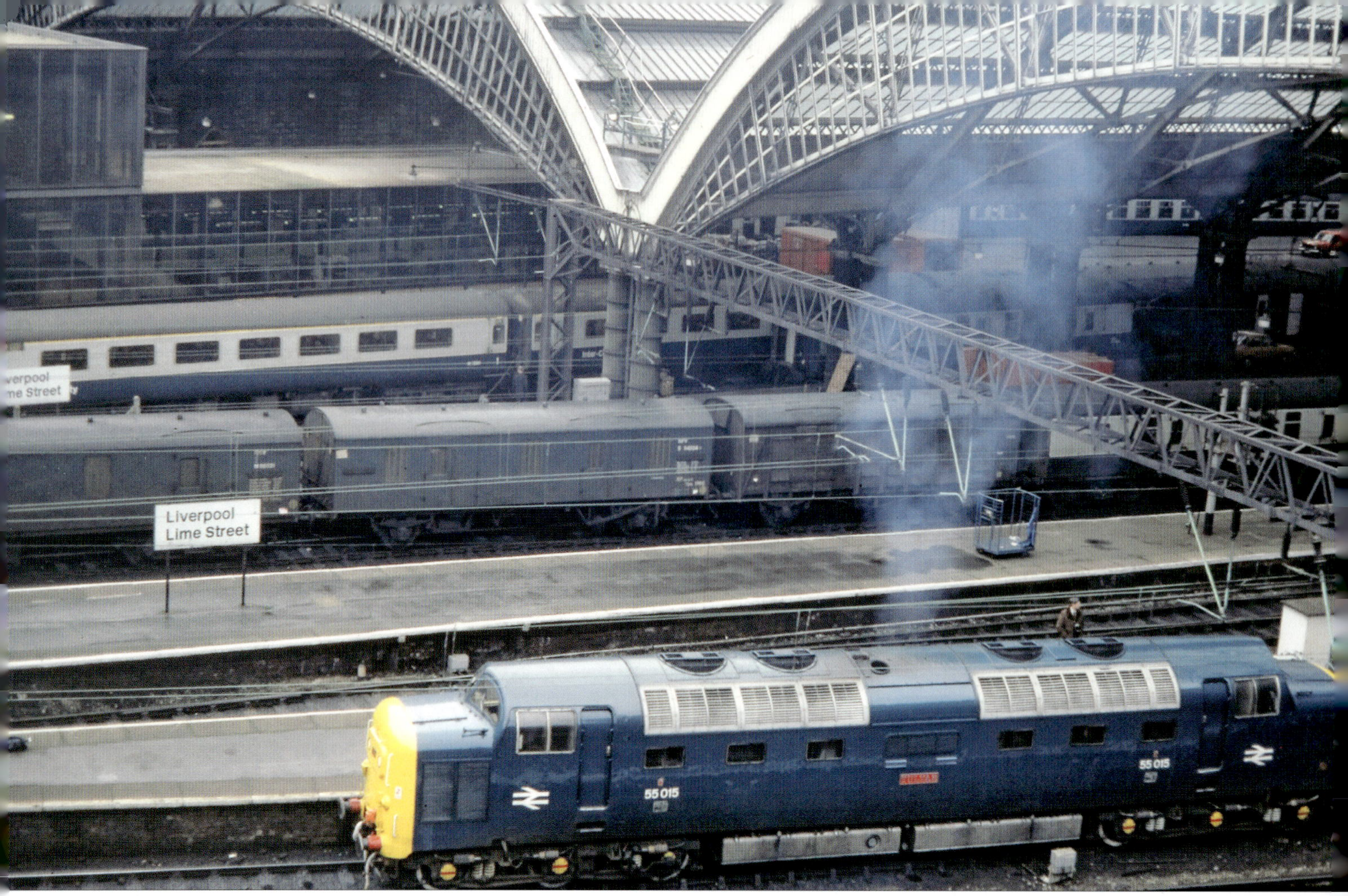
Liverpool
Lime Street
55 015
55 015

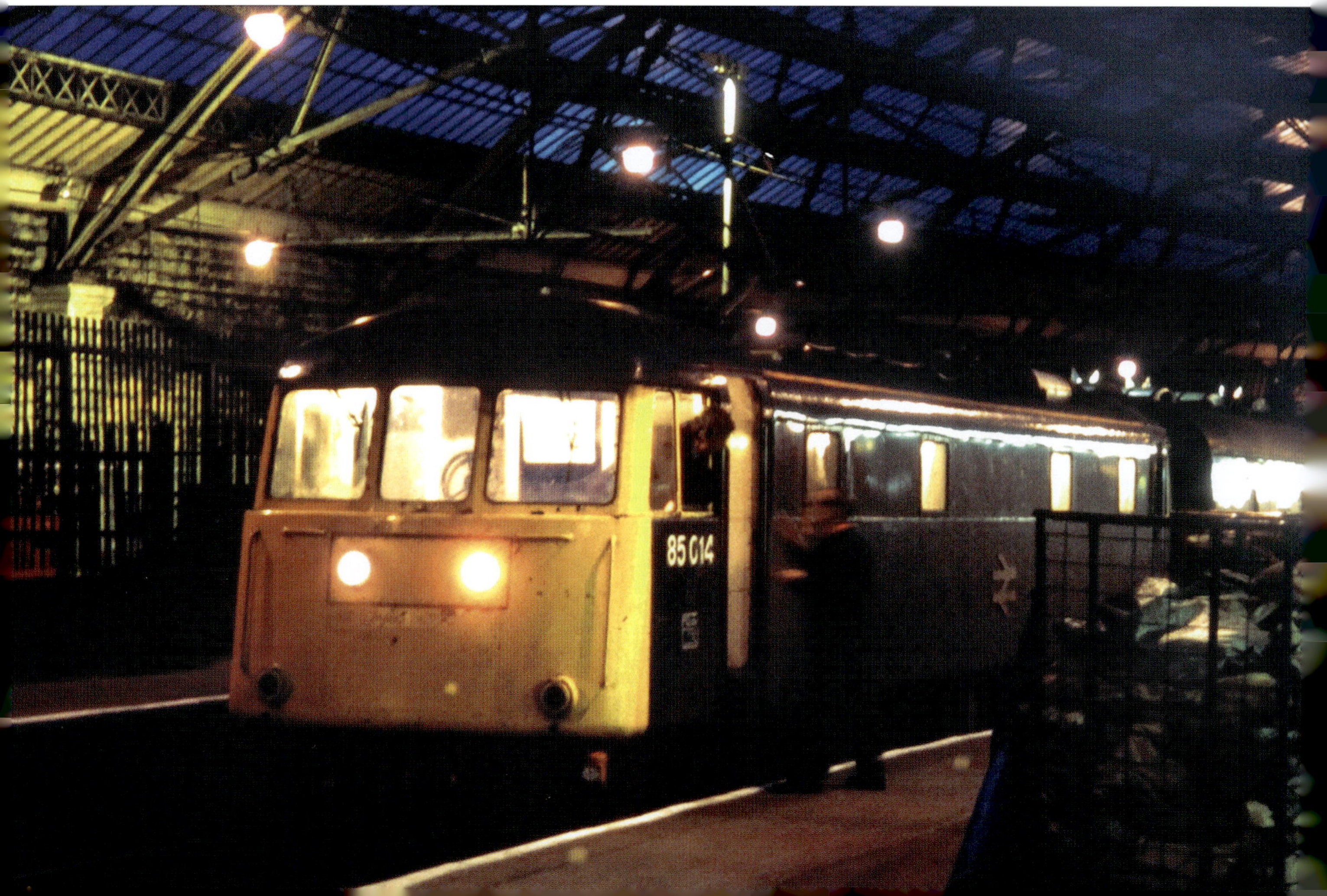
85 014

Right. A particular feature of Lime Street's train shed arches is that for no obvious reason the paired arches, each with a 212-foot span, were completed in 1879 not in a straight line street-to-exit orientation but after half-way they develop a decided curve to the north which is unrelated to the location of the station's exit point at the tunnel. There was no geological feature of the terrain on which the northern (1869) shed was built which then required the southern twin to follow suit. On the contrary, it appears that the northern had to be built first to shelter those platforms already busy and was built with the curve, anticipating (with supreme confidence) that a southern twin would indeed be built and would need the same curve to avoid Copperas Hill alongside the station. On 18th November 1981 Deltic no. 55002 *The King's Own Yorkshire Light Infantry*, built at Vulcan Foundry in 1961 and restored more or less to its original two-tone green livery, waits at platform 2 before re-joining its train for departure back to York at 13.05. Looking up, the incongruity of the northward bend of the train shed is emphasised by the southward bend of the track and platform underneath as they head for the station's exit. *IMG 1253*

Top left. The graceful curves of the northern and southern twin train sheds (Lime Street's main architectural feature) are best viewed from nearby high ground, which for this photo was a car park in Lord Nelson Street. The shorter but busier platforms are under the nearer northern arch, where on 2nd December 1981 English Electric Co-Co Deltic no. 55015 *Tulyar* is standing at platform 2 waiting to depart. The 22 Deltic class 3,300 hp diesel electric locos had dominated express services on the East Coast Main Line after the demise of the Gresley and Peppercorn 4-6-2 locos in the early/mid 1960s. Now as the Deltic locos (so named for the two 'Deltic' engines in each loco) approached withdrawal at the end of 1981 with the advent of the HST, they were in demand by enthusiasts for week-end special charter trains. To ensure that all was well with the week-end's selected loco it was the practice to use it mid-week on the normal service 08.50 York to Lime Street, returning to York on the 13.05 departure. At least three other Deltics visited Lime Street at this time, probably the first scheduled appearances by the class since the prototype DP1 *Deltic* was employed on the Merseyside Express in 1956-7. *IMG 1275*

Bottom left. . Overhead electrification of the West Coast Main Line (WCML) at 25kV AC was proposed in the British Transport Commission's (BTC's) 1955 Modernisation Plan for BR, and the first stage was completed with Crewe connected to Manchester (1960) and to Liverpool (fifteen months later). Electric multiple unit (EMU) slam-door trains of class 304 were introduced on the commuter service from Lime Street to Crewe in 1961, and through trains to Euston eventually began in 1966. To build locos for the new system in the early 1960s, the BTC agreed contracts with four private sector groups – 60 locos in all – plus 40 to be built at BR Doncaster Works. The external appearance of all five Bo-Bo classes was essentially the same following the BTC's guidelines. On 14th June 1983 class 85 no. 85014, built as AL5 E3069 in 1962 by British Rail Doncaster Works, was at work late into the night loading a parcels train on the south side of Lime Street station. In terms of performance there was little variation between the five classes but technical issues affected the availability and hence useful lives of class 84 (AL4) and to a lesser extent classes 82 and 83 (AL2/3). The arrival of 100 locos of class 86 in 1965-6, also restricted to 100mph but with improved suspension and wheels, freed the 81s and 85s to less demanding work, and a high proportion of both stayed active into the 1990s. *IMG 1729*

Merseyside and West Cheshire Railways 1965-1990

Below. Parked up at Lime Street station on 26 February 1989 are two Bo-Bo electric locos both of which are equipped with the technology of thyristor control in place of the tap-changer mechanism. Class 87/1 no. 87101 *STEPHENSON* was a unique modification of class 87/0, thirty-five of which had been built in 1973-4 to operate on the newly-electrified WCML through to Glasgow at speeds eventually to 110mph. The only class 87/1 loco was built in 1975 with its traction motor set to yield less power. It was in effect a test-bed for thyristor control that from the 1960s was also employed on ASEA's Rc series locos in use on Sweden's SJ rail network. The other loco in the picture is Class 90 no. 90006 which emerged from BREL's Crewe Works in September 1988, like the class 87s also powered by GEC traction motors. Fifty of this type were built in 1988-90 primarily to meet the demands of the WCML, releasing the less-powerful class 86 for easier routes elsewhere in the country as electrification spread. Though experience with no. 87101 seemed to attract little publicity, thyristor electronic control was evidently successful and was also employed in class 319 EMUs under construction at the same time. *IMG 1605*

Top right. On regional services from Lime Street in the diesel loco age BR relied on type 3 or the more powerful type 4 locos. Two of the latter were side by side on 30th January 1986. On the right is Sulzer-engined class 45 1Co-Co1 no. 45126 with the 14.05 service to Scarborough. Class 45 was a comparative latecomer to Lime Street after its long association with St Pancras and NE/SW services. It was developed from class 44, ten of which were built in 1959-60 supposedly for evaluation. But in the rush to eliminate steam even while new steam was still being built a further 193 were ordered of which 127 were the uprated class 45. No. 45126 was built in 1961 as D32 and withdrawn from service in 1987. On the left of the photo is class 47 Co-Co no. 47632, waiting with the 13.52 departure to Portsmouth. The Brush Traction 47s, with 512 built between 1962 and 1968, were very successful. Lighter and more powerful than others in the type 4 range, they were employed throughout BR. No. 47632 was Crewe-built in 1965 as D1652, equipped in 1985 with electrical train heating and new livery. It was renumbered again in 1990 as 47848 with an enlarged fuel tank for long distance work and electric train supply for all additional energy needs. *IMG 0796*

Bottom right. Before the recent remodelling of Lime Street station, 'platform 7' served in effect as a siding in conjunction with the station's parcels facilities. The Doric style iron columns encroached so much on platform space that safety would be jeopardised if passenger trains were accommodated. Hence on 11th August 1989 Brush type 2 A1A-A1A loco no. 31418 was on parcels duty alongside and beneath handsome recently-painted red ironwork, the line of paired columns marking where the two train sheds (northern and southern) meet. Before the TOPS computerised system of classifying locos with unique identifiers was implemented in 1971-5, this loco was D5522, built in 1959 by Brush Traction at Loughborough. With TOPS it was initially class 31 no. 31104, then in 1973 to sub-class 31/4 with electric train heating (eth) installed, as no. 31418. Withdrawn from service in 1995 by BR it was acquired for private preservation at Midland Railway - Butterley. The class 31 Brush Type 2 eventually proved itself to be one of the relative successes of BR's Pilot Scheme for replacing steam with diesel under the BTC's 1955 Modernisation Plan, but only after all 263 of the Brush locos had to be fitted in the mid-1960s with the English Electric 12SVT power unit in place of the Mirrlees JVS12T which was prone to cracking. *IMG 0660*

Merseyside and West Cheshire Railways 1965-1990

Above. During the 1980s BR relaxed its preference for plain blue livery by introducing a Large Logo variant for some of its locos, by which the 'double arrow' symbol was extended to the full height of the loco and its TOPS number was similarly magnified. Sub-classes 47/4 and 37/4 were the main beneficiaries, and one of the latter, 37431 *Sir Powys/ County of Powys* was at Lime Street on 11th August 1989 ready to depart with the 17.15 to Cardiff Central. Like most of the other 308 of the type it was built at the Vulcan Foundry works of English Electric (EEVF, the rest at EE's Robert Stephenson & Hawthorn), in 1965 as D6972. It passed into private ownership with EWS and was withdrawn from service in 1999. Visible beyond the end of the platform is the Lime Street signal box – the red-brick flat-topped bay-windowed building by the mouth of the tunnel. This was built by the LMS as a Type 13 and commissioned by BR in its first month, January 1948. The last such box operated by Network Rail, it was (I learn) a 95-lever Westinghouse Brake and Signal Style 'L' frame utilising miniature levers, special electric lever locks, lever bands and associated relay-based circuitry. The box closed in 2018 with signalling control transferred to Manchester Rail Operating Centre. *IMG 0662*

Top right. Slightly reminiscent of pre-1939 LMS Club service on Southport – Manchester trains was the survival until 1966 of through coaches between Southport and London Euston. The route from Southport left the tracks of the LC&S line to join the LNWR's Bootle Branch at Bootle Jct, passing under the LYR at Kirkdale to arrive at Edge Hill via Edge Lane Jct. With Southport departures at 08.50 and 15.05 a tank engine with two coaches (one First Class) carrying 'London – Southport' destination boards would arrive at Lime Street for the coaches to be attached to a Euston departure. (The down London express stopped at Edge Hill for the Southport coaches to be detached.) On 23rd March 1966 a Fairburn 4MT 2-6-4T no. 42233 emerged from the gloom of Lime Street tunnel with the afternoon arrival from Southport. It is seen here passing an unidentified AL6 (TOPS class 86) Bo-Bo ac electric loco which is standing at the entrance to the single-track shunt line (when was it built?) which connects after 1km to the other four lines at Brownlow Hill tunnel. Building class 86 for delivery in 1965-6 was divided between EEVF and BR Doncaster works and none were withdrawn from service before 2001. The cast alloy 'lion and wheel' emblem on the loco's side remained until the ubiquitous 'double arrow' took its place in 1967. *HF-S275*

Bottom right. The special service of through coaches from Southport to Euston came to an end on 15th April 1966, and the loco chosen to bring the curtain down was 'Black 5' 4-6-0 no. 45156 *Ayrshire Yeomanry*. It brought in the 15.05 from Southport but still with the customary two coaches. What was not customary, as a result of replacing the tank engine, was having to turn the loco ready for the return trip to Southport. Lime Street's surviving turntable was located beyond the Copperas Hill bridge by the southern retaining wall, and was called into action for the occasion. This was still two and a bit years before the end of steam on BR (an event in which both Lime Street and no. 45156 played big parts, though the latter was not used on the final day), but it would be interesting to know how often the turntable was needed in those final years. The loco received its name when new in 1936, and it is worth noting from the photo that it still retained the domeless boiler it (and some other 'Black 5s') received when built. The Ayrshire Yeomanry was a military unit whose long history from formation in 1798 to recent times defies summary. *HF-2-56-27*

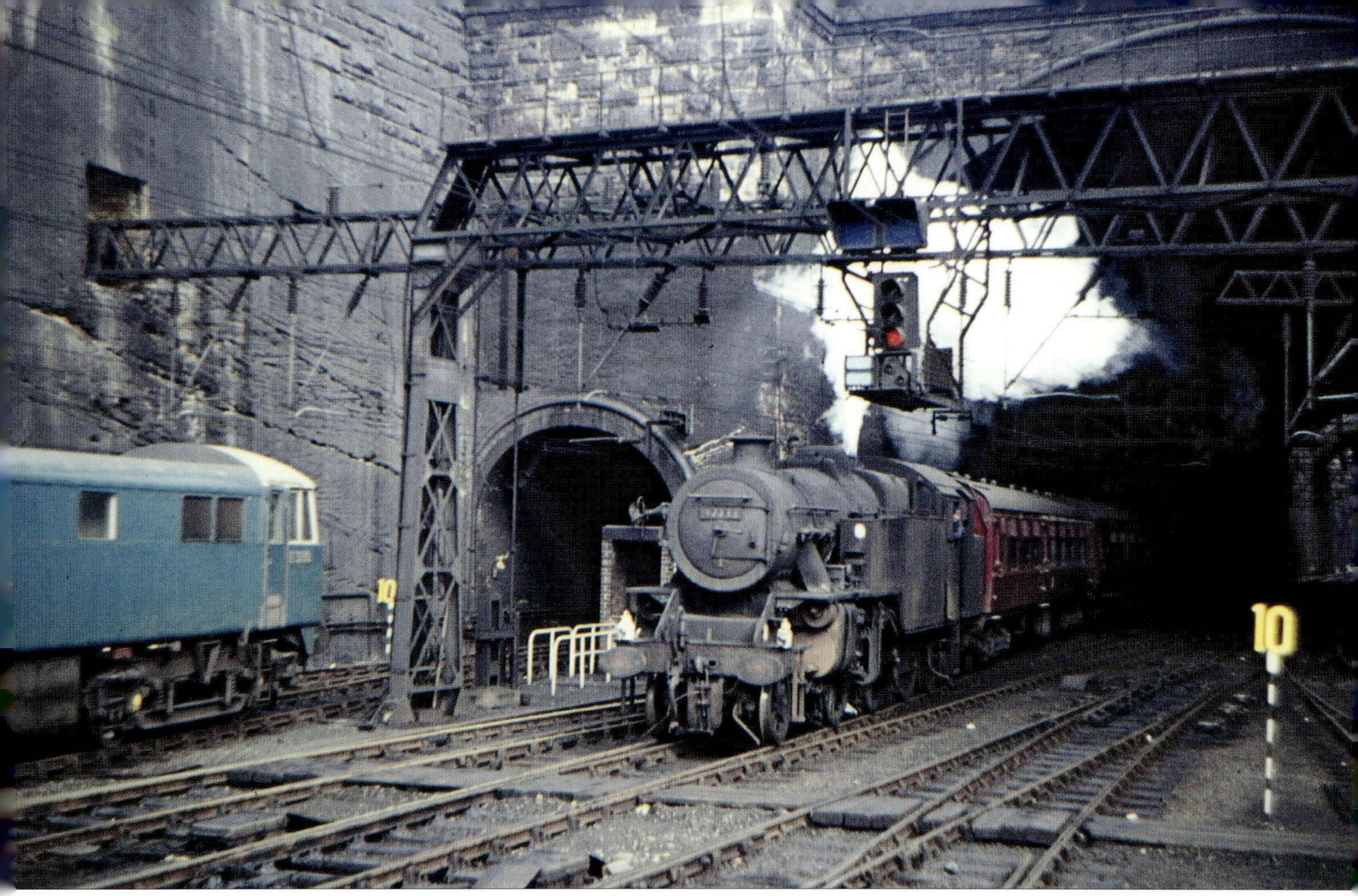
10

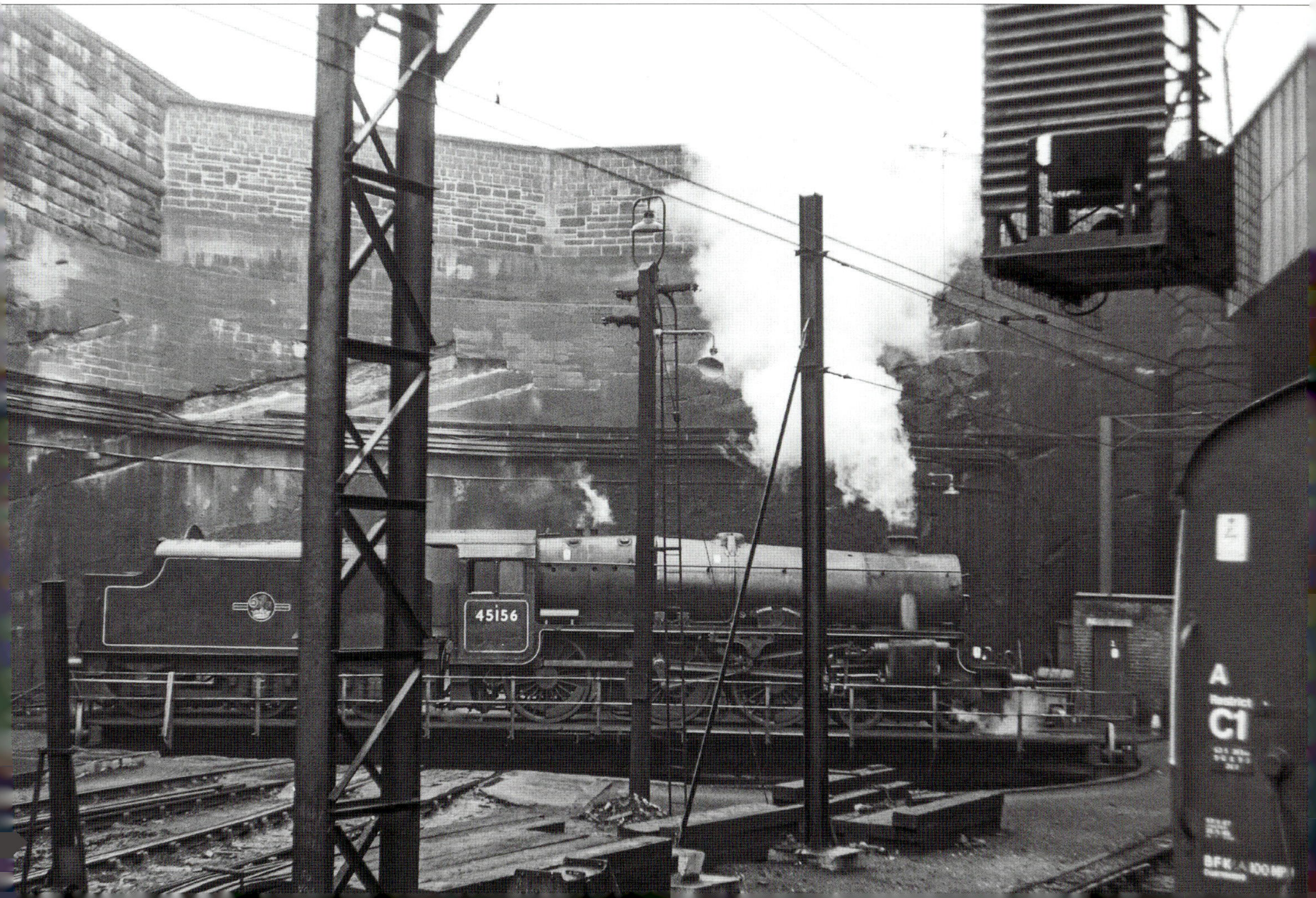
45156
A
C1

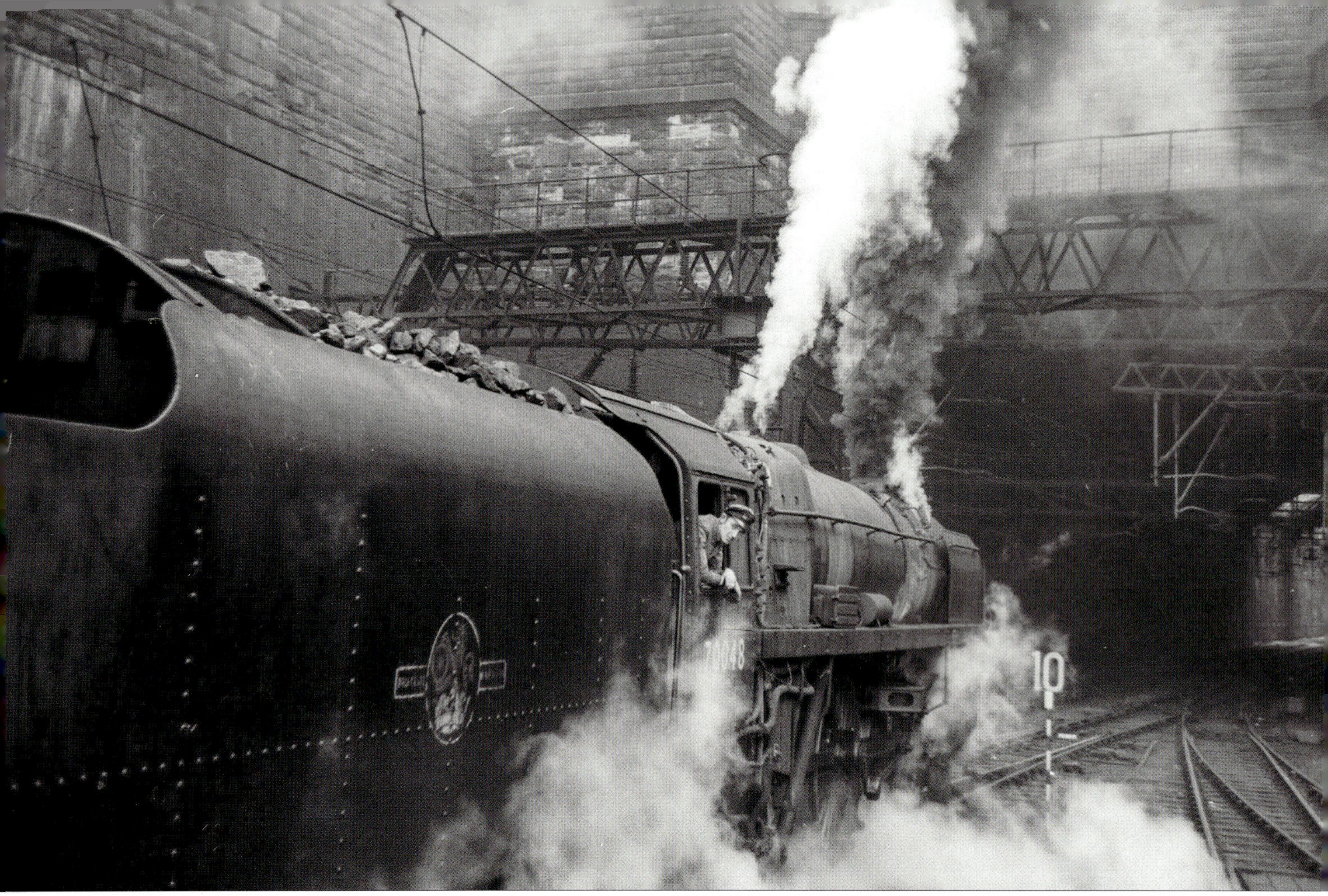
10
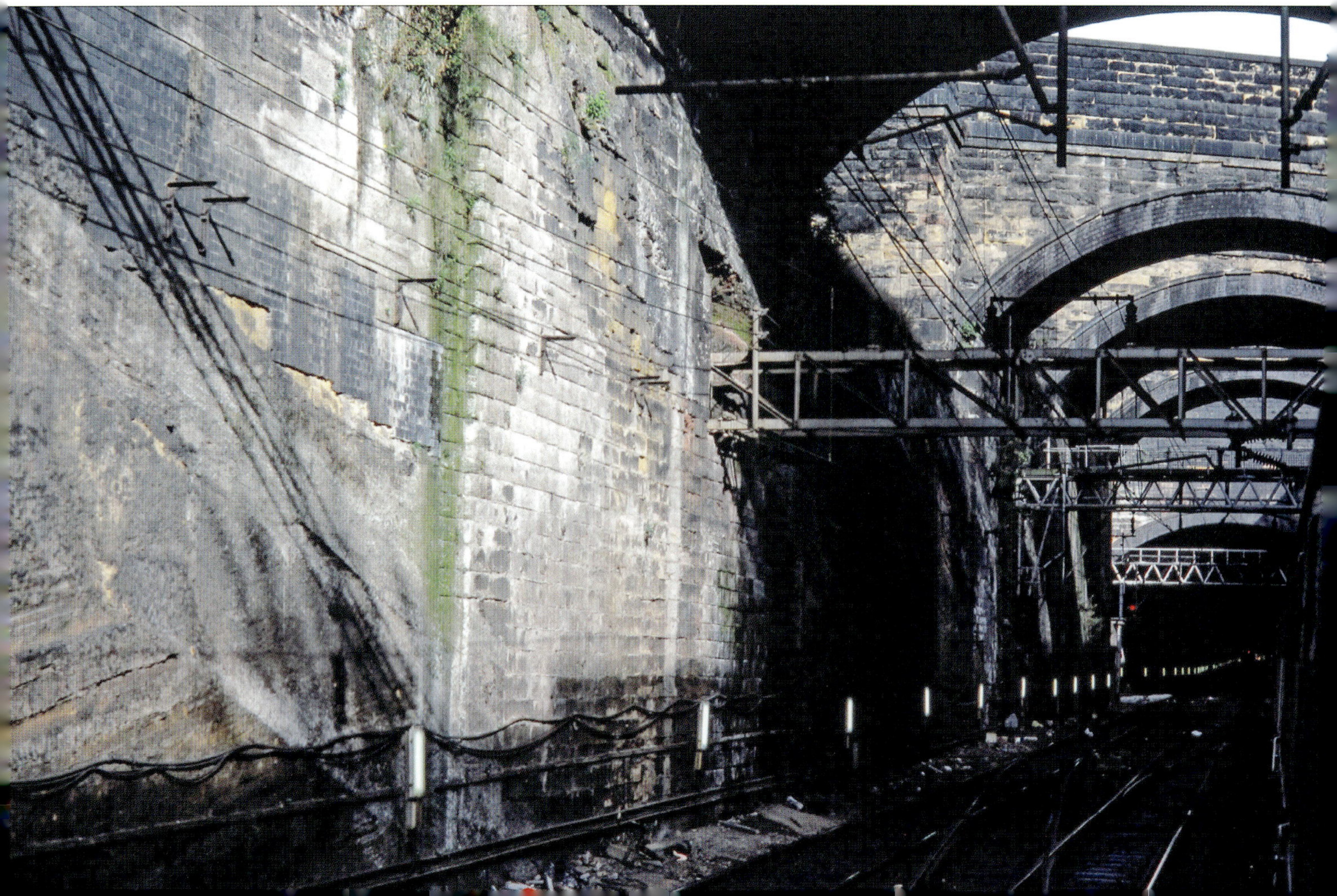

Merseyside and West Cheshire Railways 1965-1990

Top left. On Sundays in 1965 maintenance work on the overhead wires would quite often cause the power to be switched off. The obvious replacement for the electrics were the English Electric (EE) class 40 1Co-Co1 diesels, the first 'type 4' locos to be delivered during 1958-62 under BR's 1955 Modernisation Plan. Edge Hill had eleven of the 1959 batch in 1962 and still had four in 1964, primarily allocated for the boat trains between Euston and the Liverpool docks. Especially for passenger work the 40s found little favour as substitutes for the electrics though it was most probable on 23rd May 1965 that none of the diesels was available. So on this occasion attention turned to steam. There's no knowing how well Britannia class 4-6-2 no. 70048 (formerly *The Territorial Army 1908-1958*) actually performed in charge of the 17.35 to Euston, or whether perhaps it came off the train at Crewe though it had loaded enough coal for the full trip. The smoke-box door carried the 12B code for Carlisle Upperby but the crew assigned to this job would certainly have been from Edge Hill (8A) shed. The young fireman was evidently relishing the uproar of exhaust as the loco got to grips with the ascent from Lime Street. Where is that 'young' man now? *HF 1-4-18*

Bottom left. Tunnels and cuttings dug through sandstone rock have been a significant part of Liverpool's transport history. But even when the Wapping and Crown Street tunnels were excavated in the late 1820s, digging through sandstone was nothing new. In 1810-40 'King' Joseph Williamson of Edge Hill was having local rock dug out though no-one knows why. So in the depressed years after the Napoleonic Wars there was no shortage of cheap but experienced labourers to build the first two tunnels, and then to complete the new tunnel to Lime Street by 1835. For about 150 years trains exiting from the northern half of the station have had first to pass under St Vincent St/Way before all departures pass under Copperas Hill on their way to the tunnel exit. Viewed from an ascending train (seen here on 18th August 1989) the short stretch of Russell Street tunnel opens to a cutting with four street bridges before the tunnel resumes. The appearance of walls and bridges where the sandstone shows through any later coverings is dramatic with enough moisture to sustain some vegetation adding colour. No pre-cast concrete sections here. But in February 2017 pressure on part of the tunnel wall from external sources caused a major collapse and the closure of Lime Street station for a week. *IMG 0677*

Above. Approaching Edge Hill station and the summit of the climb from Lime Street, the two central tracks pass through a short tunnel underneath Tunnel Road. This section with its elegant sandstone portal is the last remnant of the original 1836 bore still in use. On 24th July 1984 an unidentified High Speed Train (HST) set forming the 12.10 service from Lime Street to Plymouth was about to pass. To its right and just visible are the twin portals of the Overbury Street tunnel. In the early 1970s BR sought a way to improve its long-distance passenger business with two projects under development. The Advanced Passenger Train (APT) offered very high speed and great potential especially for the WCML north of Crewe with electrification pending. But problems delayed the innovative APT project, while the relatively conventional prototype HST was successfully trialled in 1972-3. Production of the HST continued from 1975 to 1982. It had cars powered by Paxman Valenta diesel engines at each end of a fixed set of seven coaches. Lime Street's passengers on routes to the south-west were relatively late (1984?) to experience the benefits of its 125 mph speed and new Mk3 coaches but nationally HST sustained the passenger side of BR (and then privatised companies) for decades. *IMG 1406-1*

Above. For most students of locomotive design, the work of Stanier as Chief Mechanical Engineer (CME) of the LMS from 1932 until effectively 1944 bears comparison with the best work of any of his peers in that decade. Perhaps best known for his 8P Coronation class 4-6-2 locos which for decades handled the principal express services on the difficult Anglo-Scottish route – the WCML in modern parlance – perhaps his greatest achievement was the class 5MT 4-6-0 which became known as the 'Black 5'. They were 'Black' because green livery on the LMS and BR's LM region was reserved for passenger locos such as the Jubilee class, also a 4-6-0 type but rated as 6P5F. The Black 5, of which an astonishing 842 were built between 1934 and 1951, was universally admired and seemed to be the master of almost any job. Substituting for electric power on Sunday 30th May 1965 on the 10.10 Lime Street – Euston express, 'Black 5' no. 45466 was working hard to complete the climb from Lime Street and had just emerged from the tunnel at Edge Hill. The loco was from Speke Junction (8C) shed, and handsomely turned out. The wall on the left with fine sandstone arches flanks the downhill access ramp from Tunnel Road to the south side of Edge Hill station (platforms 3 and 4). The space behind the wall, lit by the fanlight windows, was presumably used for storage. *HF- 1-6-7*

Top right. The fine buildings and warm sandstone of Edge Hill station need bright afternoon sunshine to be seen at their best, as they were in February 1985 after a brief snow shower. A class 108 two-unit diesel is standing at platform 2, probably arriving from St Helens or Wigan. The two station buildings of the LMR date from 1836, and although they were built as replacements for the originals when Crown Street was replaced by Lime Street, they are amongst the oldest stations – perhaps the oldest – still in regular passenger use. In earlier times they have been linked by an overall roof, or a wooden bridge, or have provided shelter with platform canopies, but all that has been stripped away to reveal the near-twin classical structures. Viewed from this angle looking east they are within a little of how they appeared when built, other than an extension to the ground floor of the northern building to accommodate a larger booking office. Internally, until 1870 the tunnel to Lime Street was negotiated by gravity or rope, and each of the two buildings had to find room for an engine house. Removal of these enabled extensions for larger waiting rooms and offices. Overshadowing the northern station are other substantial buildings on its far side. They were built by the LNWR to cope with the railway's huge extension around Edge Hill resulting from Liverpool's rapid commercial growth. *IMG 1051*

Bottom right. In June 1988 class 47 no. 47503 was photographed hauling the 09.05 Lime Street to Newcastle service alongside Edge Hill's platform 1 track created in 1885 when the two-track Lime Street tunnel was converted to four. The train is about to pass three huge sandstone buildings which tower above the station itself. The first and largest of these is the engine house built by the LNWR in 1849 to power rope haulage on the newly-opened Waterloo tunnel. Commerce was growing rapidly in the northern dock system, beyond the reach of the Wapping tunnel to the south, and because of ventilation problems the Waterloo access to the docks continued to rely on rope haulage until about 1900. The growth and resulting congestion of freight traffic to and from Edge Hill required the development of a complex of new sidings and a system for sorting wagons into trains. The remarkable solution arrived at was the Edge Hill Gridiron which functioned from 1882 as a prototype marshalling yard. The complex system utilised prevailing gradients supplemented by hydraulic power from the second building on platform 1 (smaller, also with a green door) and beyond that the hydraulic accumulator tower with a 'pyramid' roof. *IMG 0507*

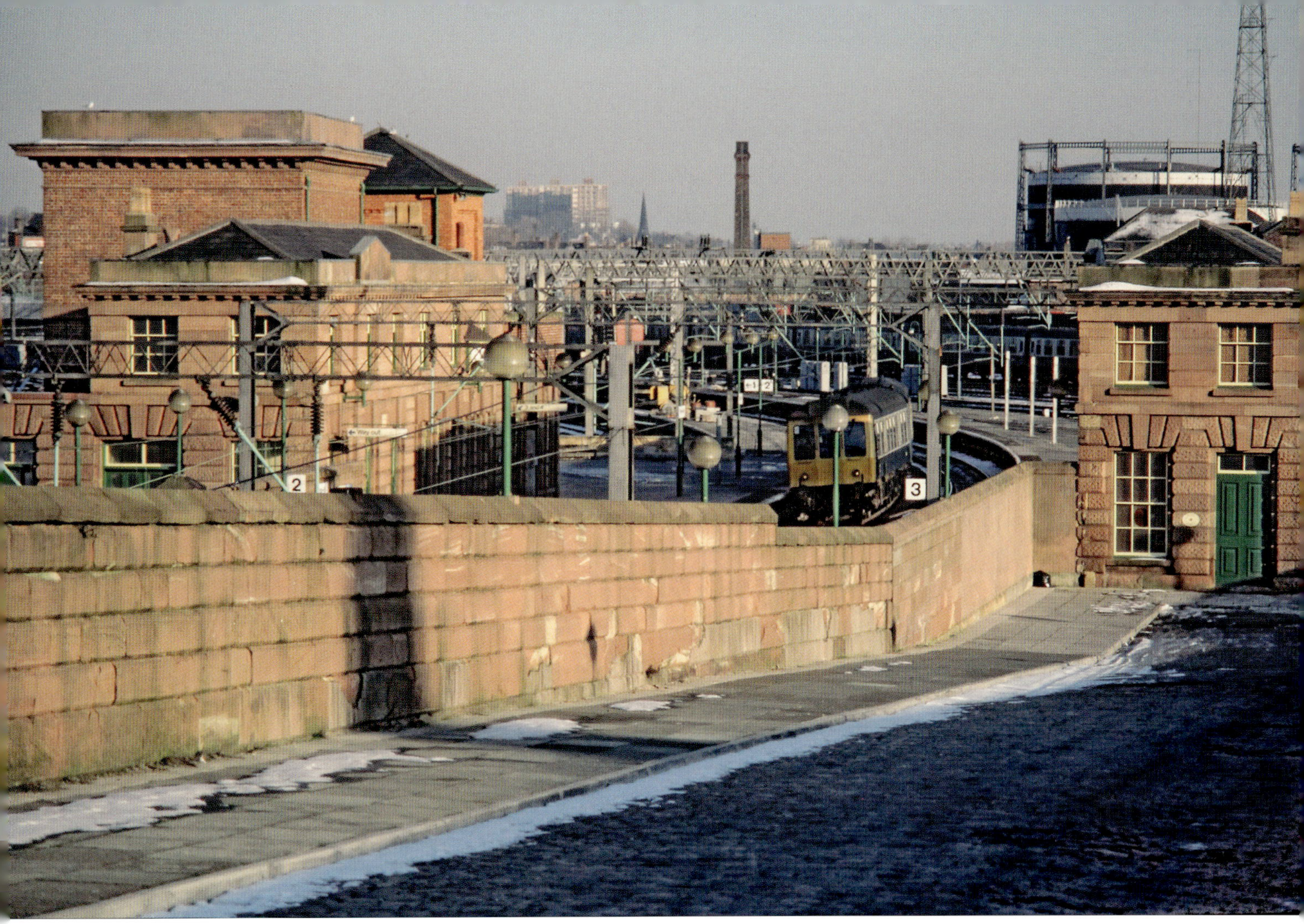
2
3

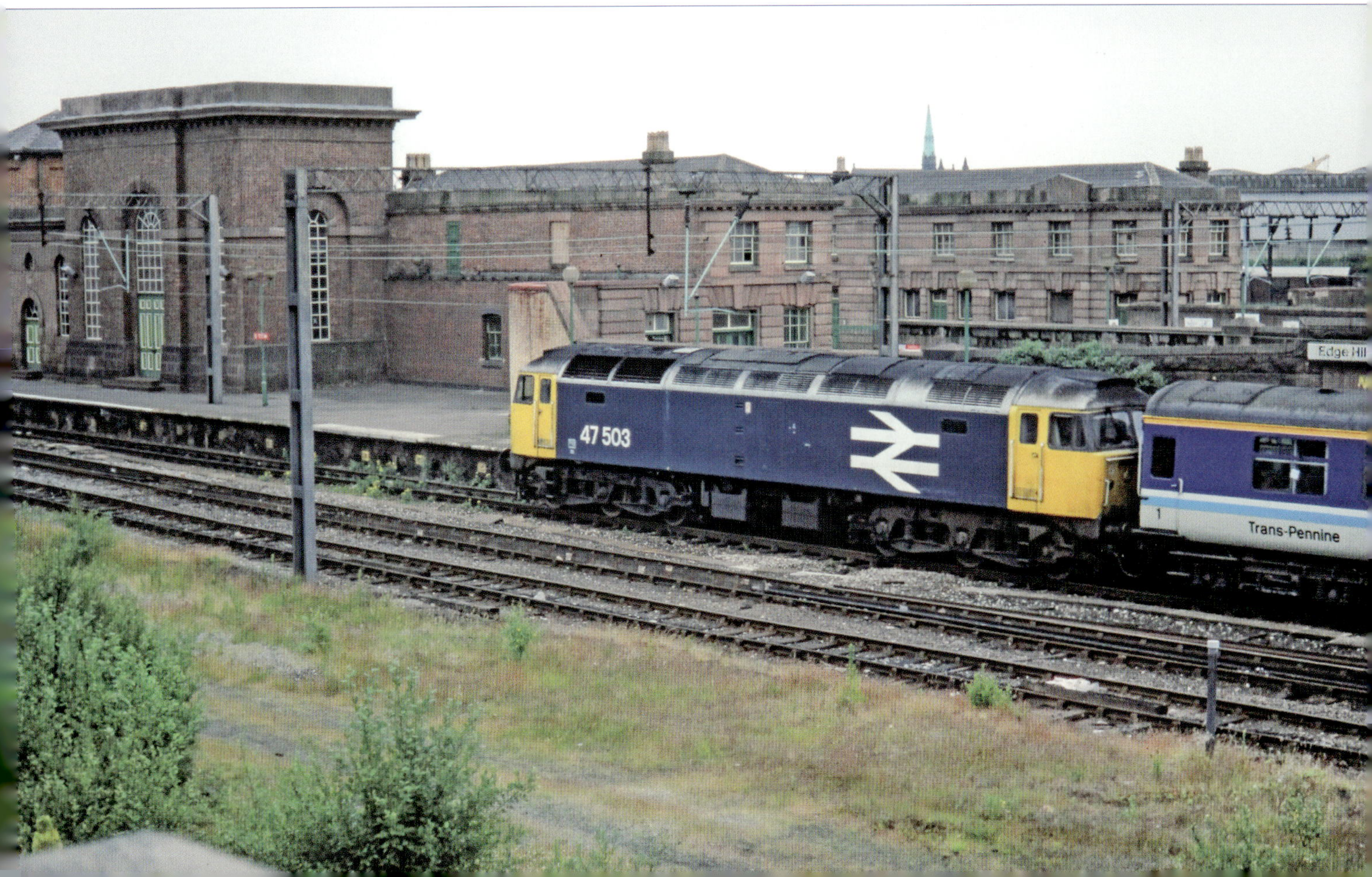
Edge Hill
47 503
Trans-Pennine

Merseyside and West Cheshire Railways 1965-1990

Top left. The more distant (and single track) of these two tunnel portals is the exit from the Lime Street tunnel to pass platform 1 of Edge Hill station. The closer portal is commonly known as the Waterloo tunnel though strictly it is the Victoria tunnel which becomes the Waterloo only beyond a short cutting at Byrom Street before arrival at Waterloo Dock Goods. The project was presented for parliamentary approval by the LMR (and GJR) in 1845 but was opened in 1849 by the LNWR. Initially the double-track tunnel was for goods traffic only, worked by cable and gravity on the Victoria section, adhesion locos on the Waterloo. In 1895 the cable failed and was not renewed so locos were employed throughout the 2 miles of tunnel. Also in 1895 passenger traffic started to Liverpool Riverside station to connect with ocean liners. Riverside closed in 1971, six months after freight traffic ended, and the tunnel closed completely in 1975. As with the Wapping tunnel, the 'Waterloo' entrance is in use only as a headshunt, which explains the manoeuvres with crewmen at each end of class 45/0 1Co-Co1 no. 45040 *The King's Shropshire Light Infantry* on 7th September 1983. The sub-class 45/0 were still equipped with steam heat boilers (unlike the electric train supply fitted to 45/1 locos) hence normally restricted to non-passenger duties as here. *IMG 1784*

Bottom left. Sunday 23rd May 1965 brought two Britannia class departures from Lime Street on a day when steam had to fill in for absent electric locos – power switched off for maintenance – and diesels – presumably nothing suitable available. In the evening no.70048 was pressed into service (p20, HF-1-4-18), but the 10.10 morning express service produced no.70001 (formerly *Lord Hurcomb*) seen here sweeping through Edge Hill station along platform 4 past carriage sidings and the concrete structure of Spekeland Depot beyond. Lord Who? Hurcomb knew about transport as a very senior civil servant pre-war who was asked by Attlee's Labour Government to be first chairman of the newly-established British Transport Commission in 1948. Labour had long believed in the merits and possibility of a centrally-organised integrated transport system, but British Road Services as a growing private industry did not have the post-war problems of Britain's railways inherited by BR. Hurcomb was ennobled in 1950, and after retirement in 1953 from the BTC (itself abolished in 1962) he became president of the RSPB and active in bird conservation. Hurcomb was succeeded at the BTC by Sir Brian Robertson in 1953 (for whom a short-lived diesel-hydraulic B-B no. D800 was named in 1958), who in turn was followed in 1961 by Dr Richard Beeching (for whom no loco was named). *HF-1-3-12*

Below. In the days when coaching stock was maroon and heated by steam, Edge Hill had a Carriage Shed in which a train's stock could be cleaned and warmed by the use of a stationary boiler, before being presented at Lime Street for the greater comfort of its passengers on the day's first departures. The Shed was just south of the station, but information on precisely when it was built and pulled down is hard to find. Early diesel locos had steam-heating boilers but too often they failed and thus caused the loco to be taken immediately out of service for repair. Electric train heating (eth) eventually solved the problem but the pattern of its installation in later diesel classes and on Mk. 1 / Mk. 2 coaches was uneven. On 14th April 1966 'Black 5' loco no. 45131 is drawing a train out of the Shed, in the period when the platform awnings were still in place and even seats were provided on Edge Hill's platform 4. The picture evokes memories of my Botanic Road days when you could get a bus into town, watch a film, and catch a late evening steam-hauled stopping train up the hill from Lime Street to alight at Edge Hill. With the window open, very atmospheric! *HF-2-54-12*

Merseyside and West Cheshire Railways 1965-1990

Below. Looking west along Edge Hill's platform 1 on 7th January 1986, a class 108 DMU is just arriving with a local train for St Helens and two passengers step lively so as not to miss it. This better view of the three huge buildings shows that the hydraulic plant in the centre is very much a scaled-down version of the 1848 engine house for the Victoria/Waterloo tunnel whose gaping portal is in the background. The accumulator tower looks to be a later add-on with its non-matching red brick and pyramid roof. The loco on the further siding (which joins its neighbour to form the head shunt) is no. 97406. Originally D335 in 1961, then 40135, it was one of four of the class 40 taken into departmental (ie non revenue-earning) stock in 1985 until the end of 1986 (and for this loco eventual preservation). The vacant land to the right of the loco used to accommodate four more sidings for goods arrivals and departures, and the support structures indicate that electrification may have been contemplated. The building high up to the right seems to be in the process of demolition. I believe it used to announce itself to Wavertree Road as LMS Fruit and Vegetable Depot. *IMG 1232*

Top right. Roughly where Wavertree Road becomes Picton Road, a bridge spans all rail lines into Edge Hill station and yard. Electrification began here in 1961, frustrating photographers with a forest of overhead line equipment which sadly ruined the classic view of trains departing from Liverpool. But on 24th May 1965 I crossed the road and was pleased with this view of Stanier 8F 2-8-0 no. 48296. From memory the loco was running light, and from the driver's searching look was held at a signal. 8Fs were commonly seen at Edge Hill, but as visitors. Given the importance of freight passing through the Gridiron it is perhaps surprising that of the 663 which were built by the LMS from 1935 onwards only five were allocated to 8A in 1951. Edge Hill shed in its prime had well over 100 engines allocated, with express locos prominent. In 1951 8A catered for 10 Jubilees (6P), 6 Patriots (6P/7P), 11 Royal Scots (7P) and 4 Princess Royals (8P). But as smaller sheds closed and express locos disappeared Edge Hill received more 8Fs, like no. 48296 which two months earlier had been reallocated from Crewe South. *HF-1-4-19*

Bottom right. The star emblem seen above on the cabside of 8F no. 48296 indicated that the loco had been fitted with wheel balances allowing it to run fast freight trains to a maximum of 50mph. I have no record of whether the 8F seen here on 15th July 1965 was similarly modified: possibly not, as this was no. 48188, pictured in front of a WD 2-8-0 and the old LNWR 1864 shed with 19 roads, re-roofed in LMS days. No. 48188 had a tragic history. On 9th February 1957 Driver John Axon drove the loco from Stockport Edgeley to Buxton, where a leak to the steam brake pipe was repaired. On the return loaded journey the pipe fractured, filling the cab with scalding steam. Axon clung to the outside of the cab to alert the Dove Holes signalman of the runaway's downward path. At Chapel-en-le-Frith station no. 48188 crashed into the rear of a freight train at an estimated 55mph killing both Axon and the guard of the other freight, For his heroism John Axon was awarded the George Cross. His story became The Ballad of John Axon, a classic Radio-Ballad for the BBC in 1958 which captivated me, then in my teens. *HF 1-24-11*

Merseyside and West Cheshire Railways 1965-1990

Above. On the 4th June 1965 Royal Scot class 4-6-0 no 46115 (formerly *Scots Guardsman)* was heading east on the Manchester line "with its next duty...the 7.50pm fast goods back to Carlisle". That description is in quotes because coincidentally John Corkill photographed the loco inside Edge Hill shed the previous day and I quote from the caption in his and Peter Hanson's book. This photo was taken from the Rathbone Road overbridge. By 1982 both the double track on the left and surviving single track on the right had been removed. The greater interest lies in the railway overbridge. The line it carried was built by the LNWR in 1887 and was integral to the Edge Hill Gridiron. It began as an off-shoot from the northbound main line near Wavertree, ran east through Engine Shed Junction, turned north over the bridge in the photo, and west over the Bootle branch to reach the Waterloo tunnel and Departure Sidings north of Edge Hill station. Visible top right is the elevated edge of the Sorting Sidings, about 30 parallel tracks in a roughly rhomboid shape. To avoid these (and the Reception Sidings further west) the line went immediately from bridge to tunnel. Since 2000 the bridge is now the site of Wavertree Technology Park station, the Park itself on land where the sidings were. *HF-1-7-10*

Top right. Time was when BR ran special trains to Aintree when the Grand National was a huge popular event. That was certainly still the case in 1967. But how such traffic from the east and south reached Aintree is debatable. Joe Brown indicates that the CLC route from Halewood Jct for Aintree race specials was not used after 1966. A less direct alternative would be the Bootle Branch to Atlantic Dock Jct, and thence via the North Mersey Branch. The photo was taken at Edge Lane Jct looking north from Pighue Lane on 8th April 1967, in gloomy fading light. The train is led by a pair of 'Black Fives', nos. 44777 and 45145, and can only have been a race special. Whatever the route to reach Aintree, if the train was now heading for home wherever, from Edge Lane Jct it had to take the Olive Mount chord (just visible beyond the hut) to reach the main line to Manchester or the WCML But it was staying on the Bootle Branch and hence may have been intending to reverse at Edge Hill for new power, or was heading for Lime Street. Was the special train's package a combined rail ticket with a night's jollification at the Adelphi Hotel? If so the celebration would have been muted. The race was won by Foinavon, a 100-1 hopeful that was so far behind the rest of the field at the 23rd fence when chaos broke out that it had an unchallenged ride to the finishing post. Footnote: the bridge behind the signal box is the famous Binns Road, the home of Meccano. *HF-S1516*

Right. Olive Mount was one of the great obstacles for the LMR to overcome by 1830. It is perhaps the most dramatic of T.T. Bury's 'Coloured Views', showing workers poised high up on the sides still hacking at the rock with pickaxes, dependent on ladders and a hoist, while a train passes below. Four miles from Liverpool, the cutting stretches for about 2 miles and its deepest point is an 80 foot (24m) drop. This view is from Mill Lane looking west on 29th June 1982, with 'Peak' class no. 45014 *The Cheshire Regiment* heading east with the 11.05 Lime Street – Scarborough service. A few minutes earlier, class 40 no. 40050 had crawled into the picture before halting beside Olive Mount Jct box to wait for the road when the Scarborough train cleared. This turned out to be a Fisons weed-killing train composed of three tankers and four carriages, with FA99900 as Spray Coach at the rear, the whole (I believe) designated Mark VI. It had emerged from the Olive Mount chord which connected westbound traffic on the LMR to the LNWR's Bootle Branch at Edge Lane Jct, which in turn opened the way to Canada and Alexandra docks. Finally, a third movement which I failed to notice at the time, an HAA coal hopper is just visible in the far left corner.
IMG 1316

Merseyside and West Cheshire Railways 1965-1990

Top left. The second view of the cutting is from Rathbone Road looking east. The subject is a two-car class 108 diesel unit forming the 13.10 local from St Helens Shaw Street to Lime Street, on 21st July 1983 – a modest train in a majestic setting. When first excavated the cutting had two tracks but widening made room for four in 1871, and the southern pair were lifted at some time between 1965 and 1982. However, this view shows that the Olive Mount Jct box also controlled a siding leading off the main line which then begets another siding, bottom right of the picture. The clue is the coal hopper barely visible in the previous photo (IMG 1316) from Mill Lane. In the late 1960s Beeching sought to rationalise domestic coal distribution by the creation of Coal Concentration Depots served by block trains of hoppers rather than wagons to isolated sidings. Fewer than 300 were built by 1970 and fewer than 100 were still active in 1985. There was certainly a CCD in Rathbone Road but date of closure is not known. A final thought: why was Olive Mount dealt with by a cutting rather than a tunnel? The LMR had experience of tunnelling before and after 1830 but little of cutting. Was Joseph Locke (who, says OS Nock, "abhorred tunnels" p.19) responsible? Certainly, posterity has benefitted from the grandeur of the cutting rather than a tunnel entrance. *IMG 1745*

Bottom left. Cronton colliery opened in 1914 as one of the latest mines to emerge on the St Helens coalfield, and was one of the last to close. (The history of Parkside Colliery at Newton-le-Willows is exceptional: opened 1960, producing 1964, closed 1993.) Located on the south side of the LMR and the M62 the tiny community of Cronton looks on the map like an outlier, about 2 miles north of Widnes. In the twelve-month period 1980-81 (the last period for which data is easily available) Cronton mine employed 572 of whom 460 worked underground, and it produced just under 300,000 tons of coal. Because of financial losses that the colliery incurred its future was reviewed in 1983 by the NCB and closure followed in March 1984. This was also the month that the NUM strike, strongest in Yorkshire, moderately observed in Lancashire, began. If there was local opposition to the closure of Cronton (as surely there was) it seems not to have left much trace. Almost a year earlier, on 28th April 1983 class 47 no. 47378 was seen leaving Cronton colliery with a train of 15 loaded coal hoppers (about 30 tons of coal per hopper), and such trains would normally have run twice a day, five days a week. *IMG 1695*

Below. The Cronton coal trains reached the Liverpool line to a point just east of Huyton Jct (where the LNWR's 1872 branch departed to St Helens) and west of Huyton Quarry station (closed 1958). About half of the 2 mile route was built as a branch in 1832 to serve two collieries then existing. It was extended just before the First World War to reach the Cronton mine when it opened in 1914. In more recent times it passed under the M57 in 1974 and traversed the M62 east of J6 in 1976. The photo shows class 47 no. 47378 on 28th April 1983 emerging at a road junction where Hale View Road meets Hall Lane. The train crosses Hall Lane, passes narrowly between two houses, and will curve slightly to join the main line. The building behind the loco is the Parish Church of St Gabriel, Huyton Quarry. The eventual destination of the train was probably no further than the Coal Concentration Depot reached from Olive Mount Jct about five miles to the west, or Fiddlers Ferry Power Station. *IMG 1873*

Merseyside and West Cheshire Railways 1965-1990

Below. It seems remarkable that Liverpool had no direct connection to St Helens until the LNW built the line from Huyton Jct on the LMR in 1871. The alternative route had been the inconvenient 1858 link line with a reversal from the LMR's St Helens Junction station. The Lancashire Union Railway (absorbed by the LNW in 1883) connecting from Wigan was not the first line from the north to reach St Helens in 1869, but it was on that line I encountered class WD 2-8-0 no. 90341 arriving at St Helens Shaw Street station with a mixed goods on 29th May 1965. The loco was a long term resident of Wakefield, carrying the 56A shed code which in 1956 displaced 25A. The photo is taken from the north end of platform 2 of the now renamed St Helens Central, underneath Corporation Road. Much else has gone since then, especially the goods depot, a fine structure of five bays (one and a half in the picture, left) and the associated sidings. Of the six chimneys visible only the square tower at the end of Shaw Street survives. What is unchanged, remarkably, is the LMS 1891 signal box, renamed from St Helens No. 2 to become St Helens Station. This follows the commissioning of the new Warrington power box in 1972 which swept away semaphore signalling and most traditional boxes in the area including St Helens No. 3. Note the ground disc signal to the left of no. 90341.
HF-1-5-01

Top right. About a mile west of Newton-le-Willows the builders of the LMR encountered the shallow river valley of the Sankey Brook. Compounding the problem was opposition from the Sankey (aka St Helens) Canal which had opened in 1757 alongside the Brook. The canal required from any bridge a minimum clearance of 60ft for the Mersey Flats sailing craft on its waters. George Stephenson's solution, the Sankey Viaduct, included built-up embankments on both sides of the valley, and consolidation of the marshy land below by wood piles. These supported the large blocks of sandstone cut from Olive Mount on which the piers of the nine arches rest. (So the decision not to tunnel Olive Mount was not Locke's alone.) The Viaduct (known locally as Nine Arches) is 200 yards long and was completed in time for the railway's opening in 1830. Grade 1 listed, the structure is celebrated as the world's first major railway viaduct. Viewing it from the south-east on a cheerless sunless 20th January 1983 a two-car dmu left Newton-le-Willows with an 11.17 departure for Lime Street. Later but in even worse light class 40 no. 40034 *Accra* crossed the viaduct with a four-carriage 10.45 train from Lime Street to Barrow. There were signs that cement had been applied to deteriorating surfaces of the piers, but the viaduct now carries the overhead equipment of an electrified line.
IMG 1812

Merseyside and West Cheshire Railways 1965-1990

Left. Neither the story of how the eminent statesman William Huskisson was fatally injured at Parkside on the grand opening day of the LMR, 15th September 1830, nor the elegant inscription on the tablet that was placed at the trackside to commemorate his life, need any repetition. It is sad that it has proved necessary to remove the original tablet and put in its place an exact replica. But huge damage was done to the original sometime between my visit on 18th December 1987 (when there were a few words scribbled with felt pen, and evidence of old minor repair to two corners) and its state in mid-summer 2001 which was halfway to disintegration. Fortunately the pieces went to the NRM at York which now has a 'restored original'. It seems that the Memorial with tablet was created in 1831 or 1833, but a photo exists of what appear to be civic dignitaries with press cameras in attendance with the caption 'Unveiling of the Huskisson Memorial 1913'. Might this have been a restoration of the original Memorial, but why 1913? *IMG 0451*

Above. In 1864 the LNWR laid the basis of its direct route from Lime Street to Crewe with a line south from Edge Hill to Allerton and beyond. Acquisition of the SHCRC's Garston -Warrington 1852 route in 1864 was fundamental to the plan, and when completed the two lines connected near the SHCRC's station at Speke (closed 1930). The LNWR was already planning a bridge over the Mersey at Runcorn, but until the bridge opened in 1869 it was restricted to its former route via Earlestown and Warrington. The Allerton line eventually developed a suburban passenger traffic to Lime Street, and this was extended to Garston (though less profitably and briefly) by the Garston Chord from Allerton East Jct in 1873. The first intermediate station to be built was Mossley Hill in 1864, followed chronologically by Wavertree (opened 1870, closed 1958) and Sefton Park (above Smithdown Road, opened 1892, closed 1960), on track that was quadrupled in 1891 throughout as far as Ditton Junction. Between Sefton Park and Mossley Hill stations the line passed under Penny Lane but at the wrong end for Beatles fans – just a hint of blue suburban skies but no other link. Viewed from the bridge on 9th May 1966 'Black 5' 4-6-0 no. 45284 was in charge of an up freight, smartly turned out and one of 8A's own. *HF-S424*

Top right. The first Mossley Hill station opened with the line in 1864, but when the line was quadrupled in 1891 a new station had to be built. Probably this involved nothing more than a relocation from the north side of Rose Lane to the south, but the station differed from the others on the line (including Allerton) in having a small goods depot and modest associated sidings that was still active until 1964. Later on the area would receive its own share of the smaller dwellings that proliferated in the early decades of the 20th century, but "In the nineteenth century Mossley Hill was probably the most exclusive residential area of Liverpool" (Liverpool Heritage Bureau p.197). Accordingly, it was the practice even after 1948 that passengers could join or alight from the principal London – Liverpool expresses at Mossley Hill by notifying Mossley Hill officials for up trains and Crewe for down trains of their wish to do so. No such privilege was needed by the lone passenger waiting to board 1D26, the 10.35 from Lime Street to Caernarfon on the very wet 13th August 1966. The loco on duty was 'Black 5' 4-6-0 no. 44866. *HF-S894*

Bottom right. The decision by the LMS to open a new station in 1939, 'West Allerton' between Mossley Hill and Allerton, reflected a growth sector in the inter-war economy which was otherwise and for many people a depressed period. But house-building prospered as did home-ownership for some families, and semi-detached middle-class dwellings spread over sizeable areas of South Liverpool. The station was located below Booker Avenue, but in Paul Anderson's well-judged words was "starkly functional" (p.23). I reckoned the end of platform 1 under the bridge was the best place to photograph class 37 no. 37430 *Cwmbrân* as it approached West Allerton with the 17.14 stopping service from Lime Street to Cardiff Central on 25th April 1989. My quarrel with the station would be equally with its meaningless name, especially as Allerton is now buried under Liverpool South Parkway. South Liverpool City Council is divided into eight wards, seven of which have or have had stations named for them. The exception is Greenbank which at least sounds attractive, but the CLC got in first with a station by that name on its Chester - Northwich line. *IMG 1652*

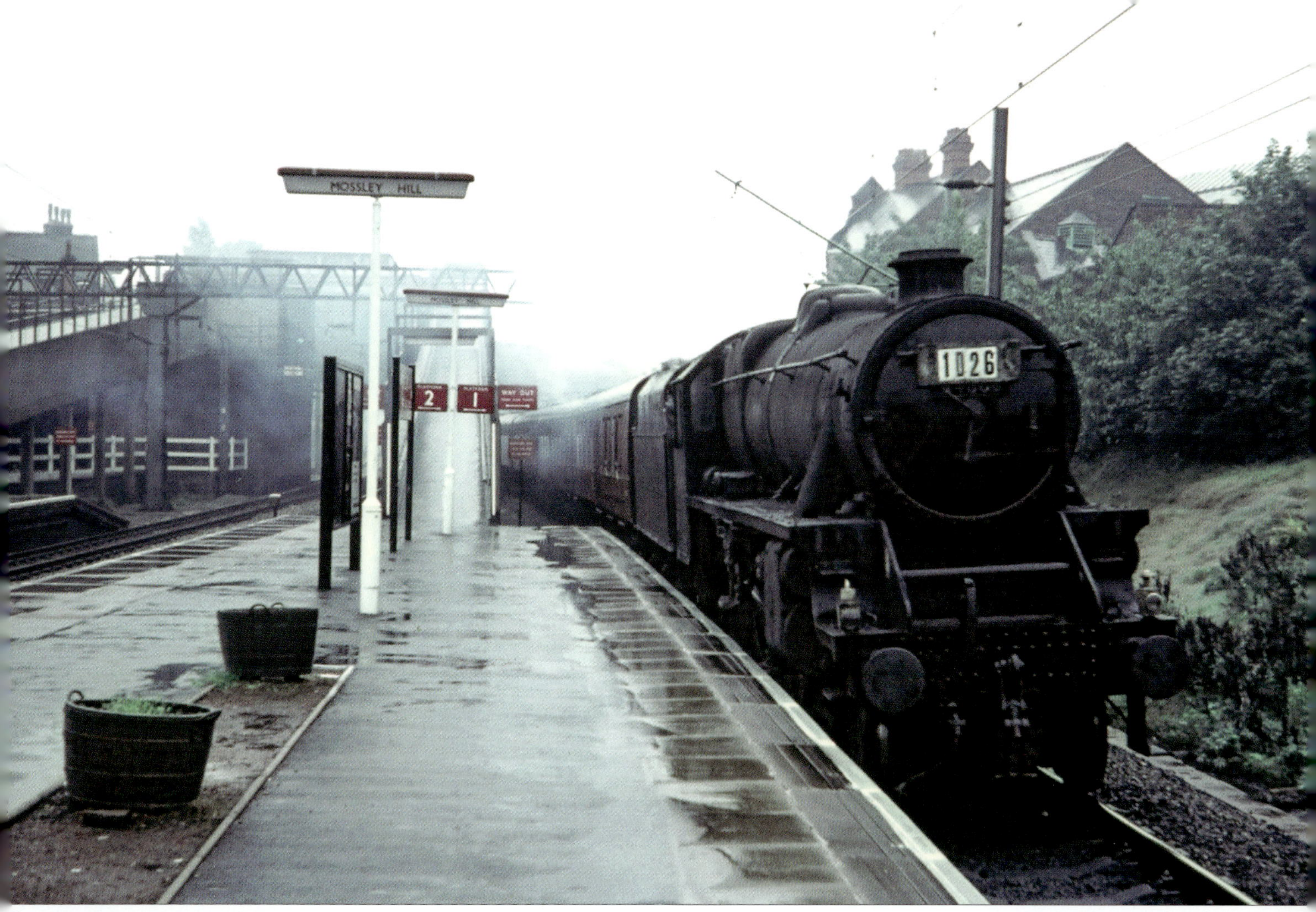
MOSSLEY HILL
1026
2
1
WAY OUT

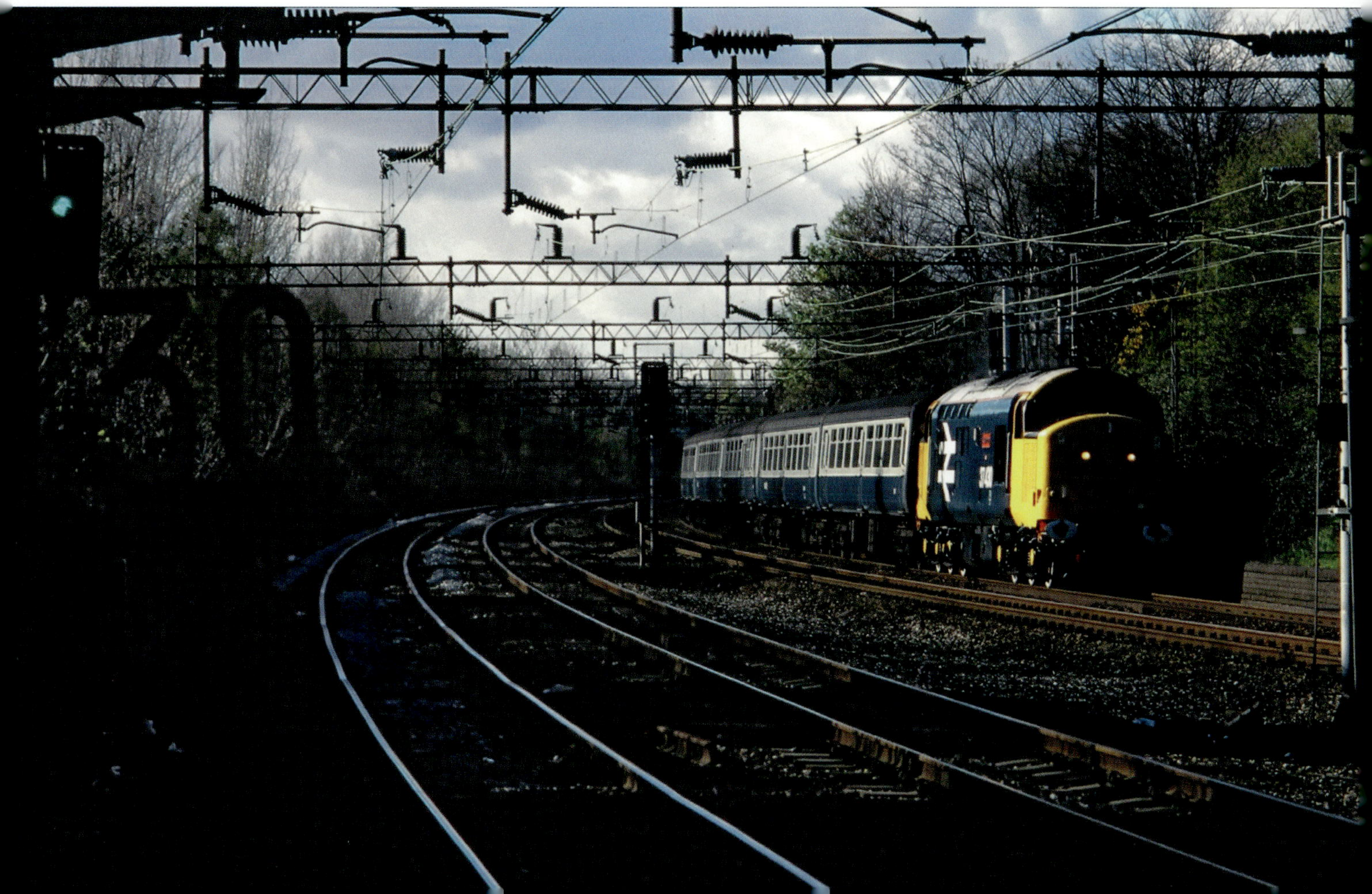
30

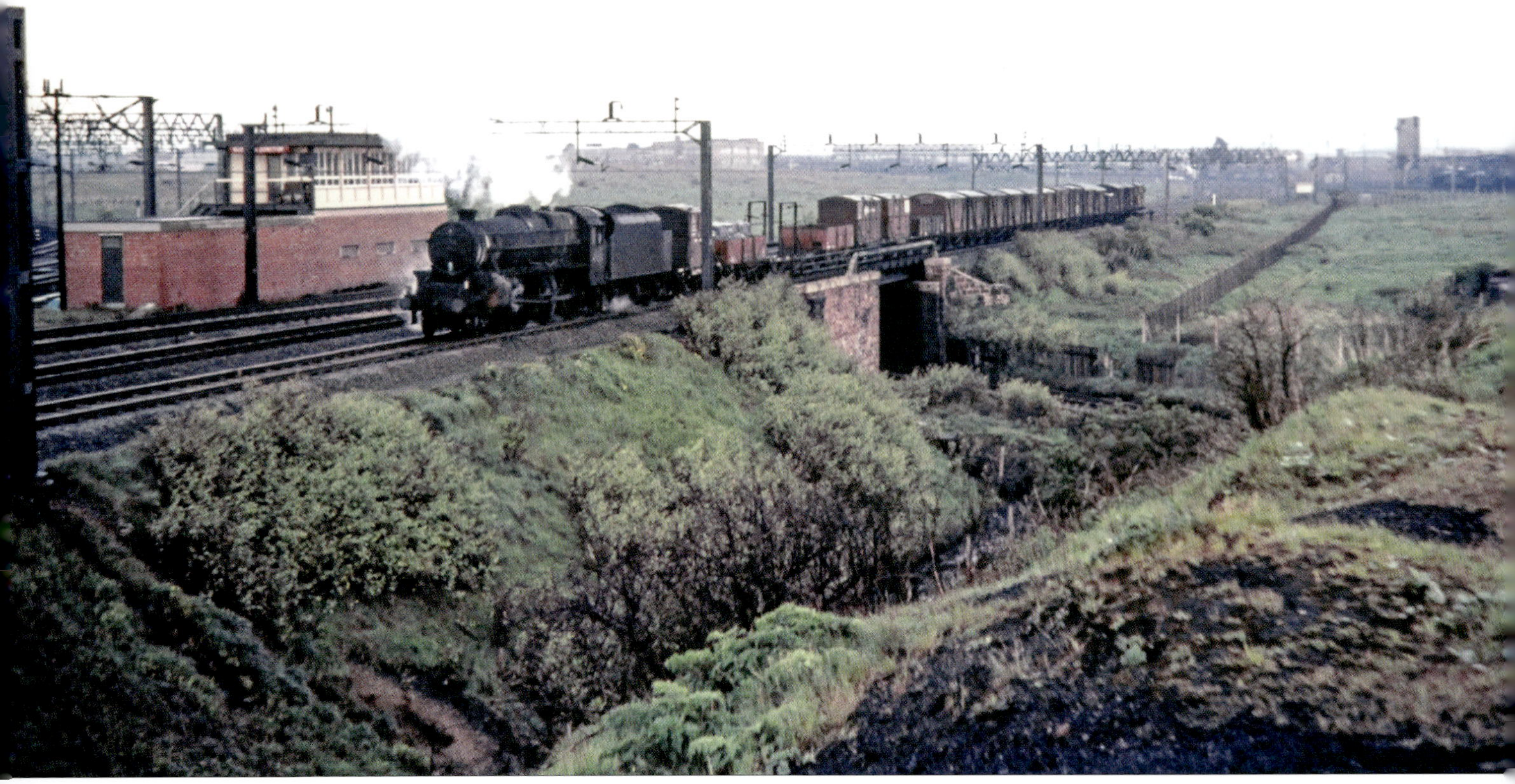

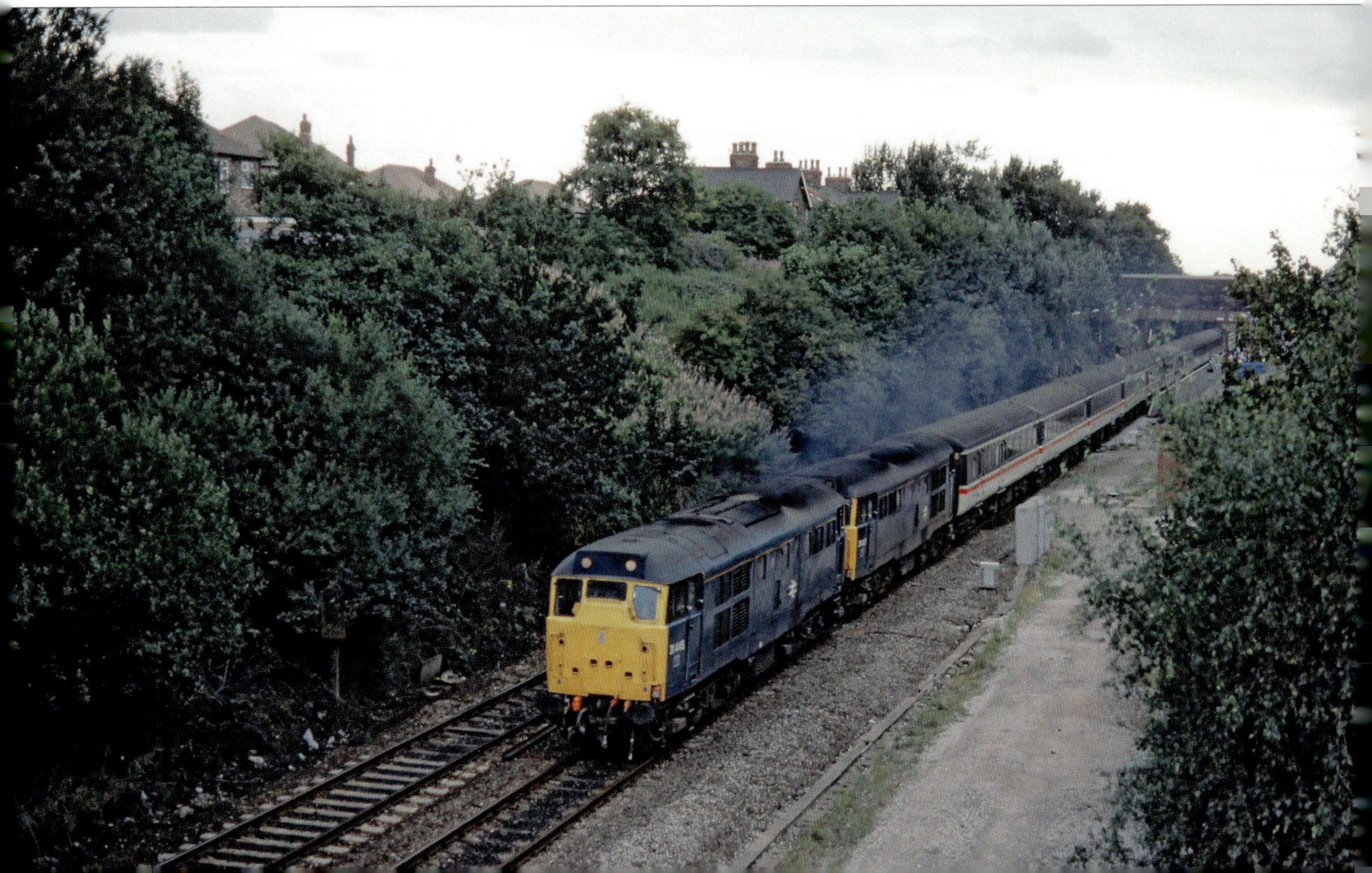

Merseyside and West Cheshire Railways 1965-1990

Top left. On 26th May 1967 I took this photo from a corner of South Liverpool FC's Holly Park ground adjacent to Allerton station platforms. 'Black 5' 4-6-0 no. 45059 is passing with freight on the down slow line and has crossed over a bridge above the tracks of the CLC's 1873 main line. The CLC lost no time in building a half-mile chord from Hunts Cross West Jct giving access to various facilities (including briefly a loco shed) but also connection to the rival LNWR's Allerton station. In 1966 this chord acquired new significance (and a new type 15 signal box for Allerton Jct), as it enabled closure of Liverpool Central High Level station and diversion to Lime Street of main-line services via Allerton to Warrington and Manchester. (Apologies that the chord lines are almost squeezed out of the picture to the left of the signal box.) In 1972 demolition of Liverpool Central and temporary abandonment of the ex-CLC track cleared the way to third-rail electrification of that track from Garston in 1978 (and Hunts Cross in 1983) to Liverpool Central Low Level as part of Merseyrail Northern Line through to Southport. Allerton station itself disappeared in 2006, rebuilt and absorbed under the roof of Liverpool South Parkway which also sheltered a relocated Garston. Final note: the huge tower on the skyline above Speke Jct merits a closer look. *HF-S1538*

Bottom left. The first station heading east from Liverpool South Parkway on the former CLC main line is Hunts Cross. This view was taken from the Mackets Lane overbridge looking west towards the lofty station which lies in a deep cutting and is, unfortunately, concealed by abundant mid-summer foliage. The visible platform alongside the passing train is an island platform, the far (hidden!) side of which is the southern terminal point of the third rail electrified Northern line of Merseyrail to Southport. Dating from CLC days the stretch of bare ground in the foreground in 1879 carried additional lines which at Hunts Cross East Jct (just beyond Mackets Lane) and Halewood West Jct peeled off to the north on the CLC's Northern Extension which until 1952 reached Southport (Lord Street). At the southern end a succession of closures, singling and freight-only restrictions ended with the last train to Gateacre in 1972 with the track lifted in 1975. Pictured is the 16.00 departure from Lime Street to Great Yarmouth with class 31/4 nos. 31445 piloting 31410 in charge. The same pair had brought in the reverse working which had departed from Yarmouth at 08.39. It is a lengthy train for a lengthy journey but surprising to see the provision of InterCity coaches rather than Regional. The date is Saturday 12th August 1989. *IMG 0666*

Below. Speke Jct loco shed (8C) was built by the LNWR in 1886 and occupied a triangular space whose straight-line base was the 1852 SHCRC line running west from Widnes to Garston. The main line was the 1864 route from Allerton heading east to Runcorn, while the sharply-curved 1873 Garston chord connected to Garston Docks. Until briefly in the early 1960s all its occupants other than shunters and for trip workings were needed for freight haulage. In 1951 these duties were shared by Stanier 8Fs, 'Crab' 2-6-0s, ex-LNWR 0-8-0s ('Super Ds'), and Stanier Moguls. By 1964 only the 8Fs remained, but were joined by 25 'Black 5s' and briefly in 1963 three 6P5F Jubilee class 4-6-0s – the shed's first 3-cyl locos – to handle car trains from the newly-opened Ford works at Halewood. Not until 1964-5 did the BR Standard 9F 2-10-0s appear, when fourteen arrived. Kenn Pearce records that no. 92091 was particularly well regarded, and very unusually had '8C' painted on the buffer beam when seen in 1966. Visited on 21st July 1966, 'Black 5' nos. 45188 and 45412 were photographed by the ash plant (built 1955 but no matching coaler) with 8F no. 48305 between them. Note the narrow-gauge tubs around the ash tower to collect ash which was tipped into a skip to be hauled up the tower and emptied into the central hopper for later disposal. *HF-3-79-28*

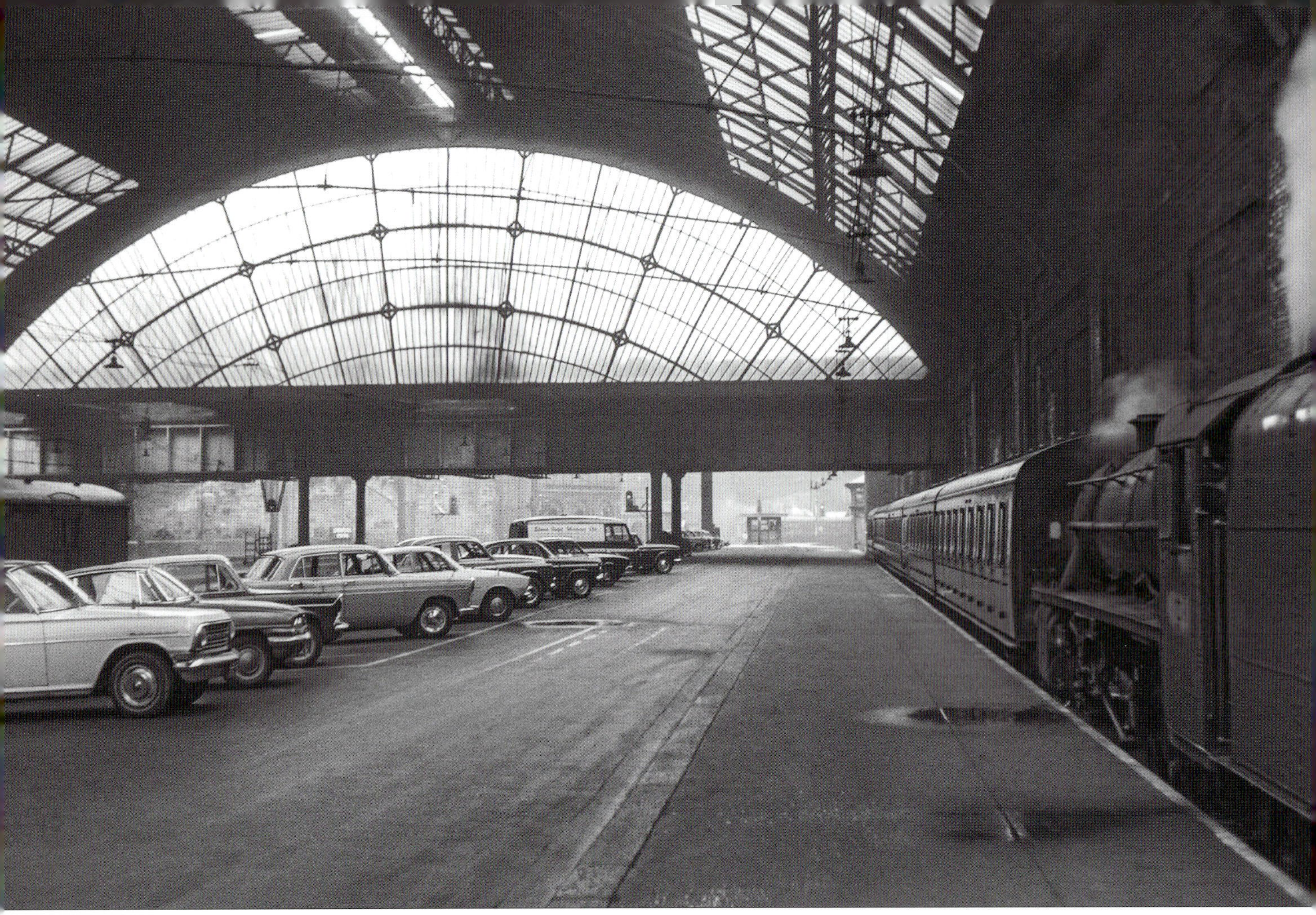

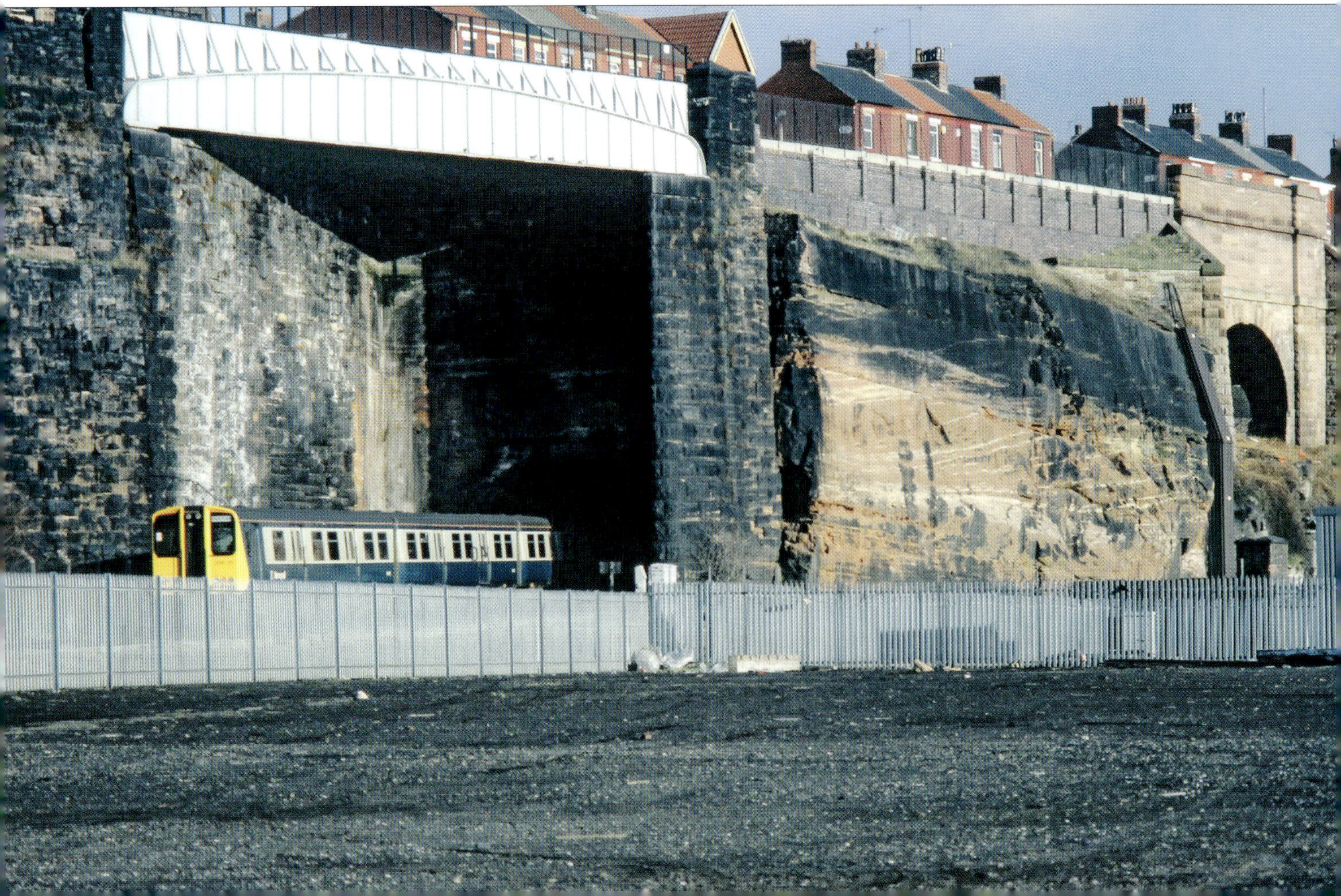

Top left. Liverpool Central was the city's third terminus to be established but it was essentially developments up-river at Garston that laid its foundation. The arrival of the St Helens Canal & Railway Co (SHCRC) in 1852 from Warrington and Widnes at Garston Dock challenged the LNWR which absorbed the line in 1864. It also encouraged in 1864 the creation of the Garston and Liverpool Railway (GLR) by joint action of the Manchester, Sheffield and Lincolnshire Railway (MSLR) and the Great Northern Railway (GNR), which built north from Garston Dock as far as Liverpool Brunswick. With the addition of the Midland Railway (thus forming the Cheshire Lines Committee) the CLC established its terminus and HQ at Liverpool Central (High Level) in 1875. On 15th April 1966, with 'Black 5' 4-6-0 no 45223 in attendance at platform 1, the station was in the same state of decline as Exchange but even further down that road: in September 1966 all services were diverted from Central to Lime Street other than the local service to Gateacre. That exemption ended in 1972 when demolition of Central (High Level) began. The height and span of the trainshed (65ft and 164ft) were smaller by one-quarter than Manchester's 1880 Central which was found a new use denied at Liverpool. But Liverpool Central's Low Level station, since 1892 the terminus of the Mersey Railway linking Liverpool to Wirral, in 1978 became an integral part of the enlarged Merseyrail network. *HF- 2-57-28*

Bottom left. On a sunny day in February 1987 a class 508 EMU is seen emerging from Dingle tunnel with a Merseyrail Northern Line train from Hunts Cross to Southport. At this point the train is still on the 1864 GLR route which it had joined at Cressington Jct, and is about to pass on its east side the cramped location of Brunswick loco shed which closed in 1961. The train then by-passes the Brunswick terminus of the GLR and runs on the CLC 1875 route towards Central station. Before getting there it runs on new track to Central Low-Level and the Link to Southport. After the closure of Central (High Level) in 1972 the line lay unused until 1978 when it was electrified at 750 V DC third rail as part of the Merseyrail system. The area in the foreground previously held railway sidings, and the infilled Herculaneum Dock. To the right of the picture is the portal to the underground section of the Liverpool Overhead Railway, inscribed "LORy: Southern Extension 1896". The Extension was built three years after the LOR opened, and its thousand-yard length terminated at a below-ground station at the end of Park Road near Dingle Lane. If the heights of the two portals look rather similar, indeed so: their subterranean tracks crossed and were separated by less than a metre, requiring some reinforcement. *IMG 1603*

Above. The oldest stations in Garston were both located on or close to Garston Way/A561, on the northern edge of the docks estate. Garston Dock station (1852) was the terminus of the SHCRC (LNWR in 1864) at Dock Road. Dock station also became the end-on southern terminus of the GLR which two founders of the CLC had taken over. The CLC had completed its long straight line from Manchester in 1873, and by joining the original GLR route at Cressington Jct (about 500 yds from Dock station) it had through running to Liverpool Brunswick and in 1875 to Liverpool Central. The CLC retained its interest in the docks with its goods depot, but built its own Garston station on its main line at Woolton Road in 1874. This station closed in 2006 to become part of Liverpool South Parkway. The LNW reinforced its position in Garston by building a Chord from Allerton in 1873 to connect Lime Street to a new station Garston Church Road less than a half-mile east from Dock station in 1881. It closed in 1939, followed by Garston Dock station and the Chord in 1947. This view of the docks, the waterfront extending to almost three quarters of a mile, was taken from Garston Way overbridge on 15th December 1983, with class 47 no. 47347 leaving with empty wagons. Partly in the picture to the left is St Michael's Parish Church, and faintly visible to the right is Garston Church Road signal box. *IMG 0582*

Merseyside and West Cheshire Railways 1965-1990

Above. Earlier on 15th December 1983 a pair of sub-class 86/3 Bo-Bo electric locos, no. 86322 piloting no. 86320, arrived with a train of containers at Garston Church Road signal box, having passed under Garston Way bridge on the far right. Remarkably the box survived until 1996, a half-century after the adjacent station closed. It was located just east (the station just west) of the bridge over Church Road with part of the structure supporting the bridge visible by the nearer ground signal. Freight traffic to the docks has declined in recent times, but the Freightliner terminal is busy.   The sub-class 86/3 locos had an odd history. Because all 100 of the class 86 locos were built in 1965-6 with axle-hung motors (rather than bogie-frame mounted) causing track damage at high speed, it was decided to fit flexi-coil springs to improve suspension. This was done reinstating the 100mph limit for most of the 86s, but nineteen (86/3) were instead given SAB resilient wheels in 1980. That proved unsuccessful so the decision was made to fit flexi-coil springs to all class 86 locos as 86/4s in 1984-5 and restore 100mph running. No. 86322 enjoyed an extended life as no. 86622 with Freightliner at 75mph, preserved at Crewe in 2021. Final note: in the photo the train loco no. 86320 has a double-arm symmetrical pantograph compared with the single arm (Z-shaped) more commonly used by BR (as on no. 86322) and other railways. Experimental or out of stock? *IMG 0581*

2 WIRRAL

Above. Birkenhead Woodside station opened in March 1878 and – regrettably but inevitably – closed in November 1967. Woodside took the place of Monks Ferry station (1844-78), the original terminus of the Chester & Birkenhead Railway. By 1860 the CBR had been absorbed by the Birkenhead, Lancashire and Cheshire Joint Railway (BL&CJR) and was jointly operated by the GWR and LNWR as the Birkenhead (Joint) (BJt). Monks Ferry was unable to deal with increasing traffic, so a new tunnel was built from Birkenhead Town station running below Chester Street for about 600 yards before opening to pass under Church Street and arrive at Woodside. The tunnel was poorly ventilated on a 1 in 93 gradient with only two tracks, imposing a bottleneck on operations made even worse by ECS movements to Green Lane sidings beyond the tunnel. The station had two semi-cylindrical cast-iron glazed train sheds supported by handsome brickwork walls and a line of slender pillars occupying the space between platforms 2 and 3. But Woodside had only five platform faces – the longest just 540 feet –with only two associated sidings. Visited on 17th May 1966 a parcels train was about to depart from platform 5 at 19.30 behind a Standard class 4MT 2-6-0 no. 76047. In the centre of the image is a Metro-Cammell class 101 DMU and directly above it is the grand station clock, by Joyce of Whitchurch. The corresponding circular space above the northern half of the station seemed to have anticipated a second clock but had decorative brickwork instead. *HF-2-81-1*

Bottom left. On a gloriously sunny 12th November 1985 class 58 Co-Co diesel electric no. 58013 is seen arriving at Garston having just passed Church Road box westbound with a train of loaded coal hoppers. Was this perhaps a grade of coal destined for export markets recovering in the aftermath of the 1984-5 miners' strike? The 50 locos of class 58 were intended to supplement the 135 of class 56 which had a poor record of availability for the haulage of heavy freight. The last twenty 56s were built at Crewe in 1983 leaving Doncaster to concentrate on the 58s during 1983-7. The main components of the 58s were the same or similar as for the 56s: Ruston Paxman engine, Brush traction motors and main alternator. The main difference was in the modular structure of the 58s with the two cabs separated in the narrow body of the loco by easily replaceable modules of power unit, electrical equipment, radiator and turbocharger. At work the 58s performed well and BR Engineering Ltd. had export markets in mind but none were forthcoming. All fifty were withdrawn early from service in the UK between 1999 and 2006. This was followed by long periods in store punctuated for many by time on hire in France (mainly) and Spain. Many like no. 58013 decayed to pieces in open store at Alizay near Rouen. *IMG 1220*

Above. Back at Woodside on 13th June 1966 I arrived just as an upper quadrant shunting signal at the platform end moved to 'off'. The loco responded with cylinder cocks open, a column of exhaust indicating that it was on the move, and an alert fireman peering ahead. It disappeared under Church Street bridge but returned to back down onto parcels vehicles at platform 5. This was 2-6-4T no. 42121, one of the post-1945 Fairburn-designed locos closely based on Stanier's model from 1935-43. On the left with a waist-coated signalman is Woodside's signal box which is hard to find in most pictures of the station. The box was built by BR in 1954 in the functional LMS style in brick, with a flat (12-inch?) concrete roof and a wide window giving a good overview of the station. Obscured by the loco is Woodside's turntable that was behind the signal box and had access to the tunnel by a separate bridge under Church Street. The impressive buildings high above the loco in Church Street (or beyond in Chester Street) will certainly have been replaced by now. *HF-3-26-10*

Top right. Evening visits to Woodside coincided with the departure of the 19.30 parcels to Crewe, the loco for which seems to have depended on what happened to be available. On the 15th August 1966 it was the turn of Stanier 'Black 5' 4-6-0 no. 45447 to draw the train out from platform 5 on a lovely sunlit evening. Just by the loco are two trolleys, one of which seems to have a load of coal but surely not 'Loco' in August. To the left are a pair of DMUs, a class 101 on platform 4 but on the siding is a much less common variety, thought to be class 120. On the other side of the wall is the roadway down from Rose Brae to what was intended to be the grand entrance to the station. Beyond that (to the south) are the graving docks and towering cranes of the Grayson Rollo and Clover (later Western Ship Repairers) shipyard. Ironically, it seems that in 1961 Woodside may have ceded some of its space alongside the station to the shipyard, but about fifteen years after the station's closure the shipyard also closed down with the graving docks filled in. *HF-4-8-23*

Bottom right. The memory of Woodside I cherish most was the fine view to be had from the spacious area between platforms 3 and 4, with the clock tower of Birkenhead Town Hall (1887, tower partly rebuilt after fire in 1901) prominent on the skyline. To the right was the Woodside Hotel of 1834. The 5th March 1967 saw a number of special trains to commemorate the end of the through service to Paddington, with the locos involved receiving an informal clean with dabs of red paint for number plates. One such in the photo was Stanier 2-6-4T no. 42616, but the EE type 4 1Co-Co1 diesel-electric no. D297 (later class 40 no. 40097) on platform 1 received no special treatment or interest, and in those days was simply an unwelcome interloper. I believe it had worked in with the last sleeper train from Paddington (from Chester) and ended its day taking out the last through service. Passenger access to Woodside station was originally by a road bridge (Rose Brae) which crossed over the platforms between the train sheds and Church Street bridge, making this view impossible. Old photos show how dark and dismal Rose Brae bridge made Woodside, with the bridge little higher than loading gauge clearance. *HF-5-1-15*

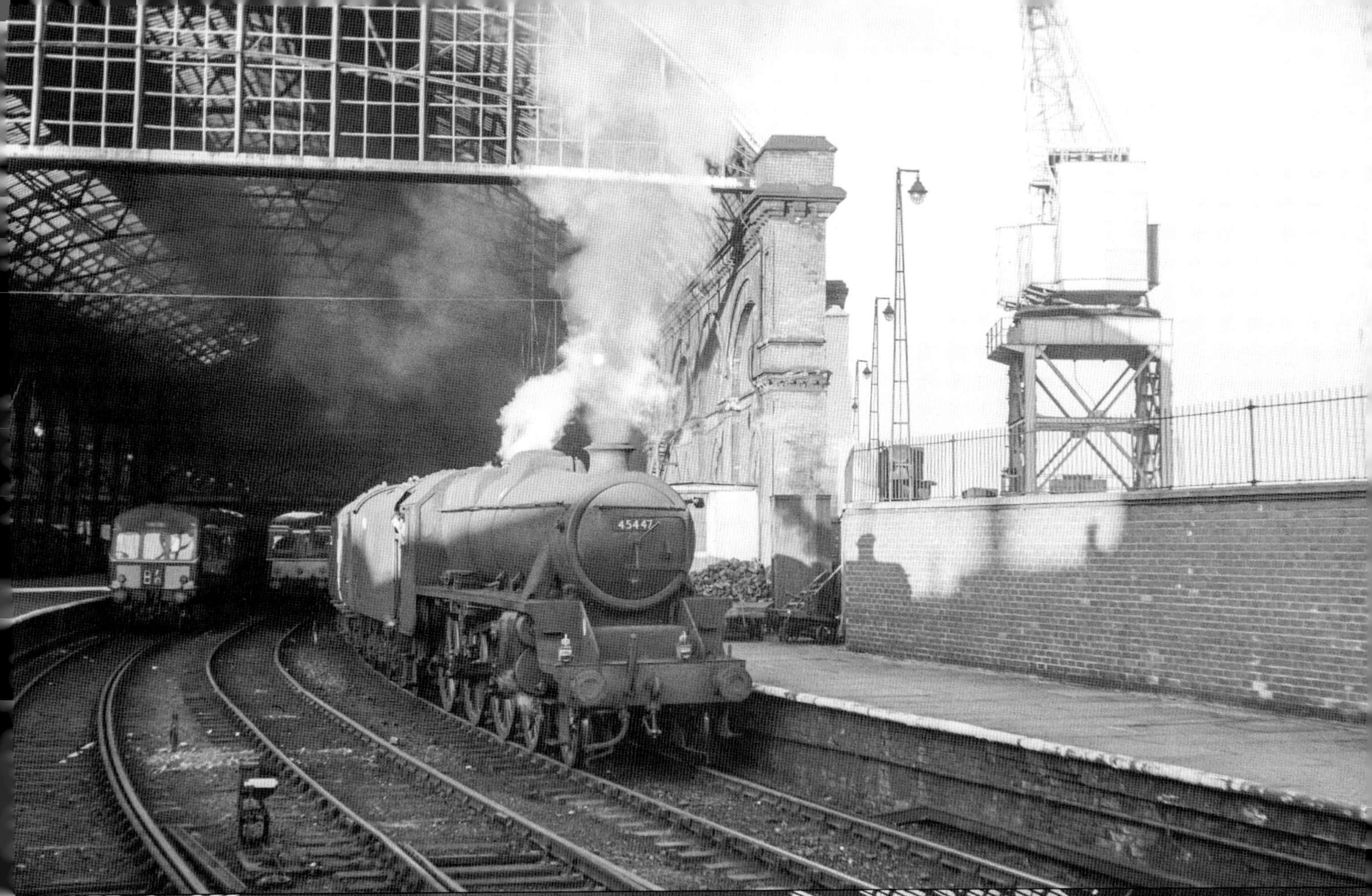
45447

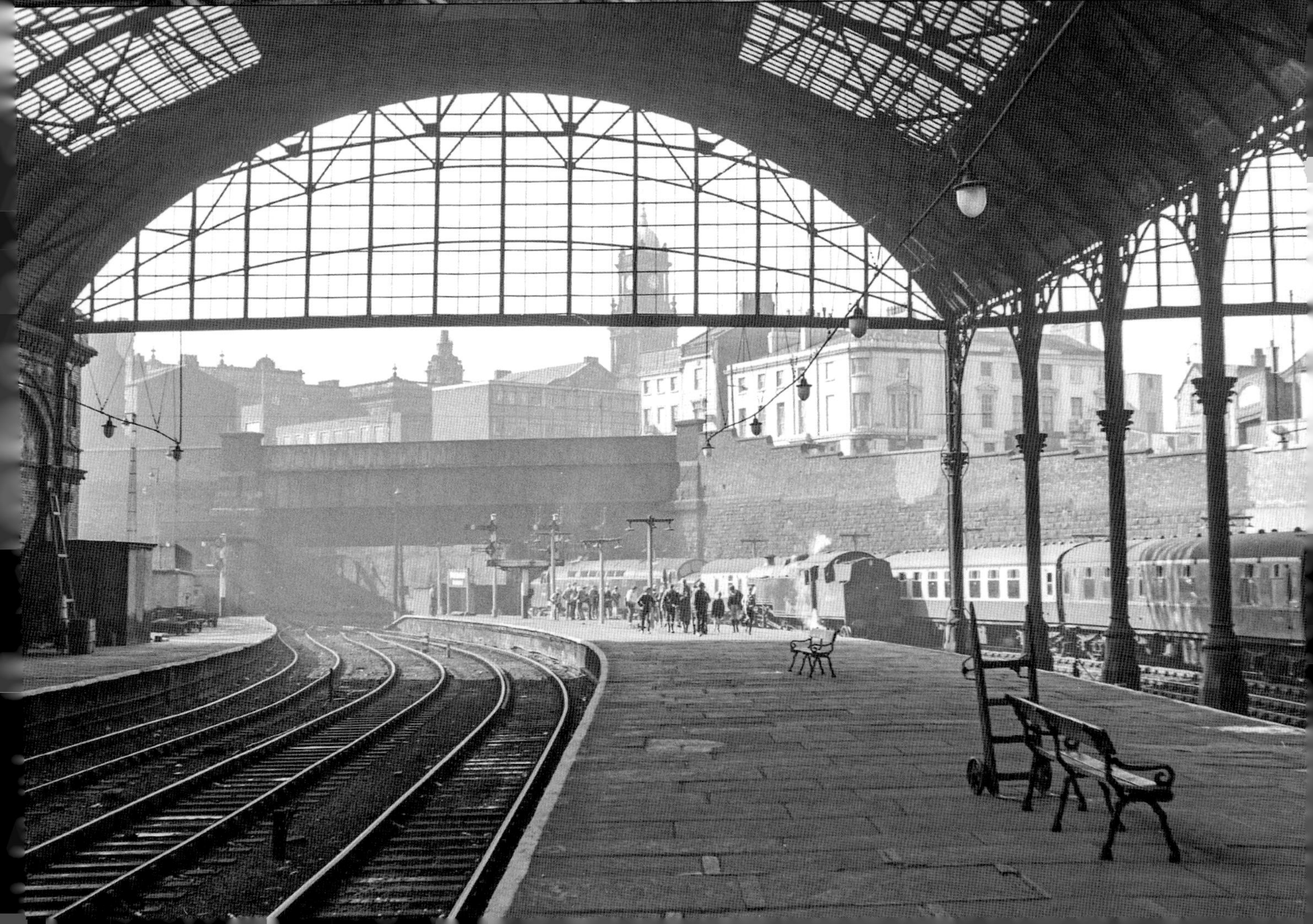

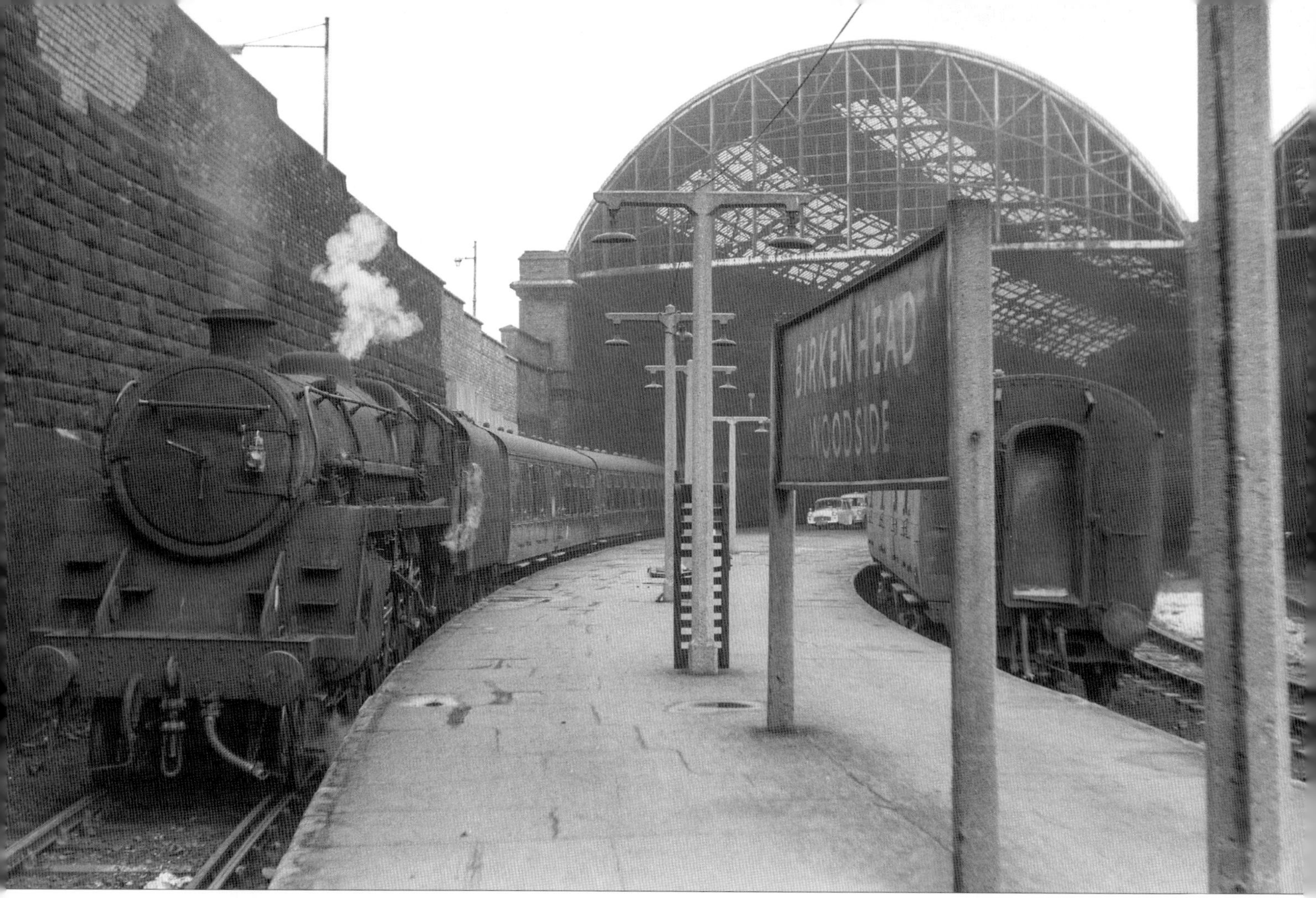
BIRKEN HEAD
WOODSIDE

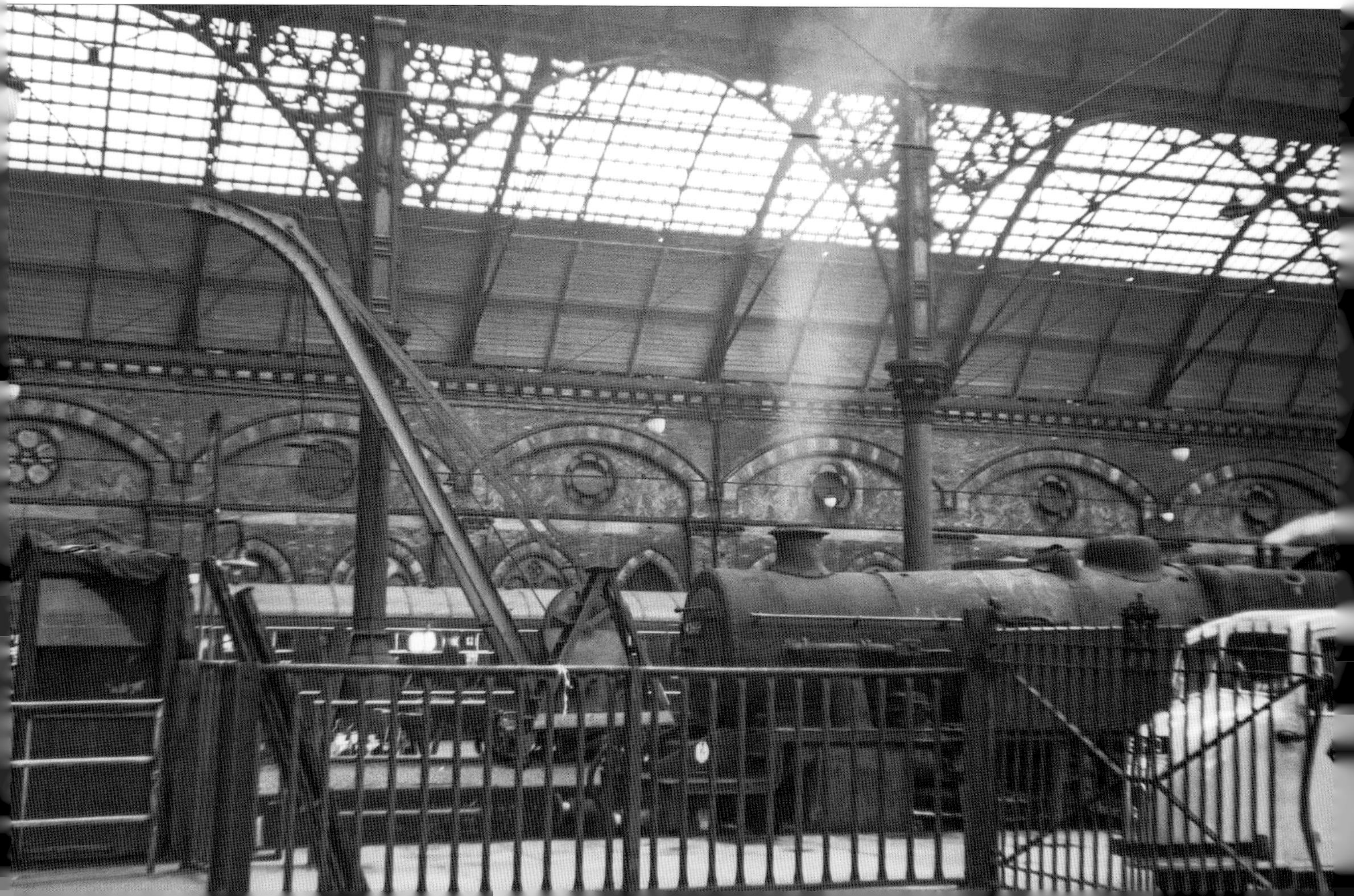

Top left. Is it only old locospotters who regret the removal of a smoke box door number plate? In this case there was no problem in identifying Standard class 5 4-6-0 no. 73094 from its cabside number, waiting at platform 1 with the 14.45 service to Paddington on 13th June 1966. Woodside was famous for being back to front: the plan was to link the station directly from its south side to the Woodside Ferry, but Ferry intransigence, tramway companies which made their terminus on the north side of the station, and the unalterable location of the longest platform (1), all foiled the plan. Instead of passengers bound for London, the elegant porte-cochère on the station's south-east corner shielded nothing more exotic than parcels trolleys and was removed along with the Rose Brae bridge, reportedly in 1961. Five years later, it is remarkable how completely the wretched bridge with the charming name had been removed leaving no trace. Why did it take so long to get rid of it, just six years before the station closed? Footnote: the car on the platform looks like a three-wheeler – coincidence, or the Scarab driver's own motor? *HF-3-27-13*

Bottom left. This view of Woodside on 17th May 1966 was taken by the customary entrance to the station on the northern side, looking south. I was attracted by the patterns in the brickwork at the back of the scene and the central columns, as well as indications of the different activities going on. Furthest from the camera is part of a DMU, probably class 101 as used on services to Chester and Helsby. There is a casual collection of lanterns at the foot of the nearest pillar. This side of the pillars is a small lightweight crane, with no obvious source of power. Adjacent to the crane is another Stanier 2-6-4T no. 42613 which has evidently arrived at platform 1. In the bottom right-hand corner is the front end of a Scammell Scarab, a 3-wheel tractor unit which when coupled to a trailer was employed by BR for urban parcel delivery services. It is in the yellow livery which began to displace maroon and cream in 1963. It bears the red 'flattened hexagon' emblem incorporating a black 'arrow' which was briefly also used to publicise the Conflat door to door container freight service. In front of the Scarab is an open gateway, and in the bottom left corner of the photo is perhaps an out-of-use ticket collector's booth and a gateway onto the platforms. *HF-2-82-3*

Above. In June 1966 Mal Pratt, then a passed cleaner, transferred from Bank Hall shed to Birkenhead's Mollington Street in search of more opportunity as a railwayman. This he found with rapid advancement to booked fireman, and passed to drive steam by early 1967. His reaction to Mollington Street (8H since 1963) as someone accustomed to life at a steam locomotive depot, even if smaller and less frantic at Bank Hall, is worth quoting: "First impressions about the shed were how dirty it was with locomotives all over the place". As an amateur occasional shed visitor, my first reaction on arriving at Mollington Road on the evening of 17th May 1966, a few weeks or even days before Mal P, was exactly as he describes his. I was taken aback by the intense level of activity and density of smoke in the choking atmosphere, all grit and grime. The preparation of locos for the day's work at a busy shed would normally go on more or less continuously, but that evening at Birkenhead it seemed that night-time freight was what 8H existed for. MP reports that in June 1966 the shed had no fewer than 56 Standard 9F 2-10-0s (compared with 'only' ten two years earlier). That was a quite exceptional concentration of heavyweight power of which three in the photo were lined up like warriors preparing to go out and do nocturnal battle. *HF-2-85-18*

Above. At the other end of the scale but also in triplicate and in steam on 17th May 1966 were three of Mollington Street's 'Jinty' 3F 0-6-0T shunting locos. These were (L to R) nos. 47533, 47447, and 47324, similarly preparing for the night's work. The depot's allocation of the shunters had fallen from eight to six in the previous two years, and another was lost that month. About 420 'Jinties' were built by the LMS during 1924-31 after the Grouping of railway companies in 1923, but the type was essentially a development by Fowler of Johnson's Midland Railway equivalent. They were not classic dock locos with a short wheelbase for tight corners but nonetheless much of their work was done at Birkenhead docks as well as shed pilot duties. Birkenhead shed closed to steam in November 1966 and the place of the 'Jinties' in the docks was taken by diesel-mechanical shunters (later numbered as class 03) built by BR at Swindon and Doncaster. The new shed behind the row of three was presumably the two-road building constructed in 1951 taking the place of half of the old LNWR depot. *HF-2-84-15*

Top right. The Birkenhead Railway (BJt), a shortened name adopted in 1860 for the (1859) BL&CJR, was jointly owned and operated by the LNWR and GWR. In 1878, needing better facilities for their respective locos, the two companies built their own separate sheds on the same site at Mollington Street. The two sheds were unalike in appearance but each had eight roads and its own turntable, Even after 1948 when the sheds became the responsibility of the LM region the division persisted to a degree until remaining GW locos at the shed were eventually repatriated. The traction requirements of Woodside were met, but goods traffic was the main preoccupation. What freight locos were available at Birkenhead? Mal Pratt tells us that 'Crab' 2-6-0 locos like no. 42782 seen here raising steam on 17th May 1966 (alongside the wall of the 1951 shed showing signs of enlargement) had extended lives because the only other large locos were the 56 Standard 9Fs which were restricted on some lines and sidings. What is surprising in MP's 8H loco list for June 1966 is the complete absence of Stanier's 'Black 5' 4-6-0s or 8F 2-8-0s. Between May 1964 and June 1966 Birkenhead lost all 19 of its 8Fs and 6 of its 11 'Crabs' but received an additional 46 9Fs. In January 1967 the last of the 'Crabs' anywhere were withdrawn from service at Birkenhead. Stanier loco arrivals saw the shed through to closure to steam in November 1967. *HF-2-82-4*

Bottom right. Compared with Walschaerts gear and piston valves which look relatively straightforward and robust, the action of the Caprotti system appears less overt and rather spindly to this amateur. The merits of poppet valves and the rotary cam shaft have long been accepted but in Britain adoption never quite went beyond the stage of trial. The LMS equipped ten Claughtons in 1928 but results were inconclusive. Twenty of the post-war 'Black 5s' (nos. 44738-57) were built with Caprotti gear in 1948, and two more (nos. 44686/7) in 1951 at Horwich with a revised version known as British Caprotti. It was this system that was employed in the unique 1954 Standard class 8P 3-cylinder 4-6-2 no. 71000 *Duke of Gloucester* whose disappointing results in service have been traced in preservation to original errors in both design and construction. 'British Caprotti' was also installed when BR decided to build thirty of the Standard class 5MT 4-6-0s (73125-54) as Caprottis in 1956. Either for specialist skills or spare parts, Caprotti locos were normally allocated to a small number of sheds: in 1951 the twenty 'Black 5s' were equally shared by Llandudno Jct, Bristol, Holbeck and Longsight (the latter also receiving the 1951 pair). With their number halved by 1964, Southport and Speke Jct then had five each, Longsight the other, while the Standards went to Derby and St Rollox, and to Patricroft (including no. 73144 seen here on 10th August 1966 at Mollington Street looking well cared-for but withdrawn three weeks later). *HF-4-5-7*

Merseyside and West Cheshire Railways 1965-1990

Above. When visited in May 1966 the three monstrous ferro-concrete towers at Birkenhead MPD were barely visible through the steam and smoke of three 9F 2-10-0s preparing to leave the shed. By common consent the 250 9Fs built during 1954-60 were the most successful of BR's eleven Standard types (and the last to appear). Their period of service was brief, and their dominance at Birkenhead limited to barely three years. On a later visit, on 10th August 1966, 9F no. 92151 was alongside the two ash-disposal plants and the wagon-hoist coaling tower. To the right of the 9F some of the 'narrow-gauge' tubs are visible which collected the ash from under the locos and were then emptied into a skip. The skip was then hauled up a tower for the ash to be tipped into the hopper, which in the image is in process at the left-hand tower. It is testament to the amount of work done by Birkenhead locos that three towers were deemed necessary, even if provided belatedly. The third tower is the coaler, a wagon-hoist type (or No. 2) planned by the LMS in its modernisation programme in 1933, but, like the ash plants, not installed at Birkenhead until 1955 (one of the few by BR, and indeed the last). In front is a Stanier two-cylinder 2-6-4T no. 42548, just re-allocated to 8H the previous week. *HF-4-6-14*

Top right On 5th March 1966 class 3FT 0-6-0T ('Jinty' shunter) no. 47447 was found on the Birkenhead Docks access line (aka the Cross-Dock route) along Corporation Road, close to the point where the BJt's ownership gave way to the Mersey Docks & Harbour Board. The depressing background scene is one of demolition. The three sidings parallel to the C-D route of the MDHB were still in place. However, the sidings which led away from the BJt line at a right angle to serve the Cathcart Street Goods Depot of the LNWR had been lifted a few years earlier. The area in the foreground of the photo was flattened to create a lorry park, and it appears that the signal post was relocated to the west to allow this development. In the background the process of demolition reveals the exteriors and interiors of some of the fine warehouse buildings being reduced to rubble. There is no sign of a shunting signal to allow the manoeuvre but the loco proceeded at least as far as Canning Street North box. The purpose of the movement was hard to determine but seemed to involve the canister (oil or paraffin?) sat on the running board. Along with Birkenhead's other 'Jintys' no. 47447 was withdrawn at the end of the year.*HF-2-25-27*

Bottom right. After the Jinty 0-6-0Ts were retired their place on the dock system was taken by Gardner-engined 0-6-0 diesel-mechanical shunters, 400 of which were built by BR between 1958 and 1962. Largely because of the changing nature of rail freight with BR's rational preference for trainload services rather than wagonloads, there was less need for small shunters, and a quarter of the 400 were withdrawn before they could receive their TOPS 03 class numbers in the early 1970s. Only a handful survived into the 1980s but three of Birkenhead's allocation, nos. 03073/162/70, were still active until they were the last to go in 1989. No. 03170 was seen on 5th July 1984 between the footbridge and Canning Street North box with what looked like a very unusual burden. A coach of any kind (other than an enthusiasts' excursion) would have no role on the dock system, but this turned out to be a Departmental vehicle, DB977100, marked as Condemned but not in fact broken up until 1995. The literature suggests that this may have been a barrier coach employed to mediate between non-matching coupler types, presumably in exceptional circumstances, but I have no idea what they might have been in this case. A year later it was at Clapham Junction so evidently it was more than just a regional asset. Notice Liverpool's Liver Building above the signal box roof. *IMG 1391*

ENTRANCE TO DOCKS 206-213
MORPETH EGERTON
WALLASEY AND
MILL QUAYS
CORPORATION ROAD
03162
CANNING STREET
NORTH

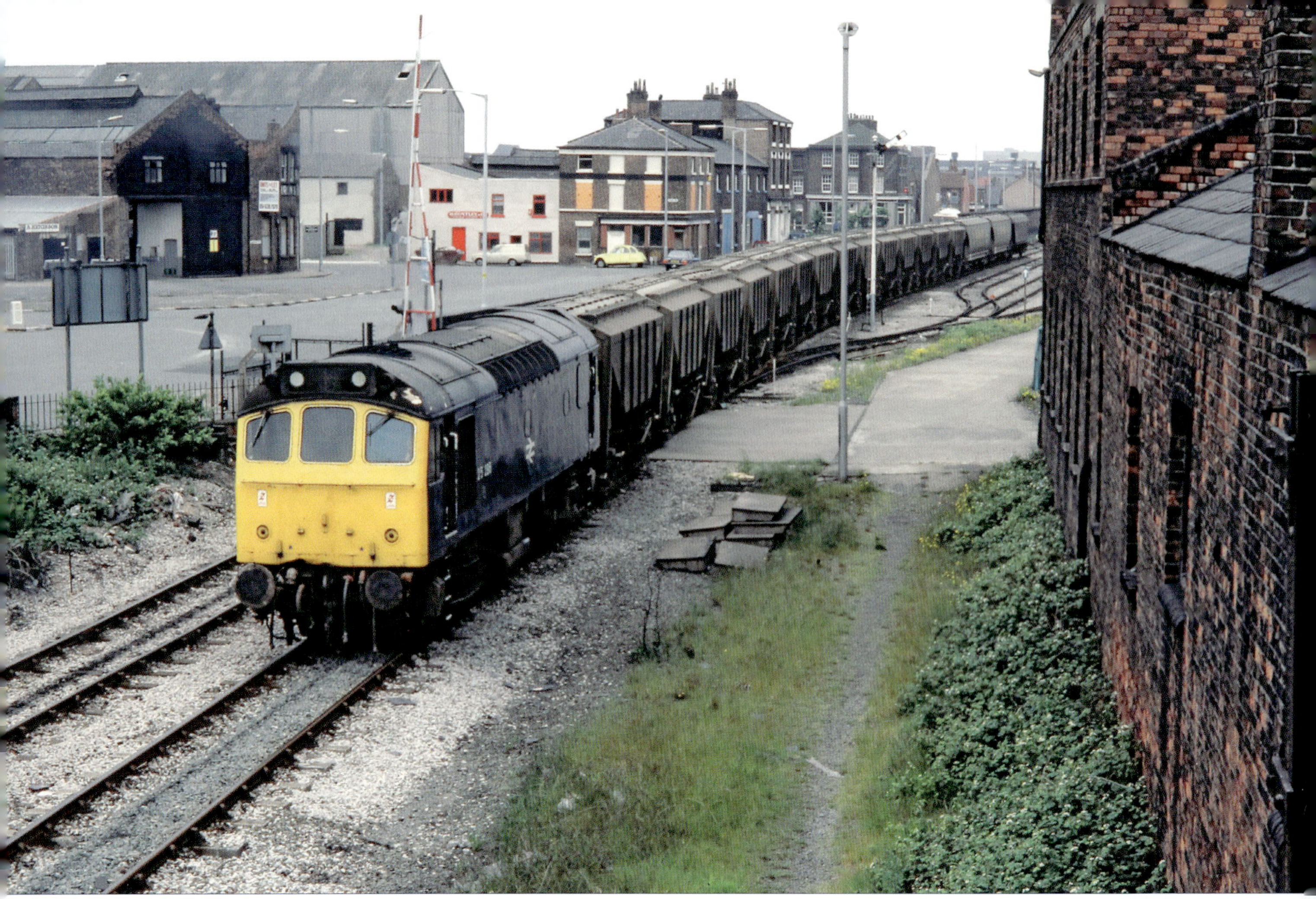

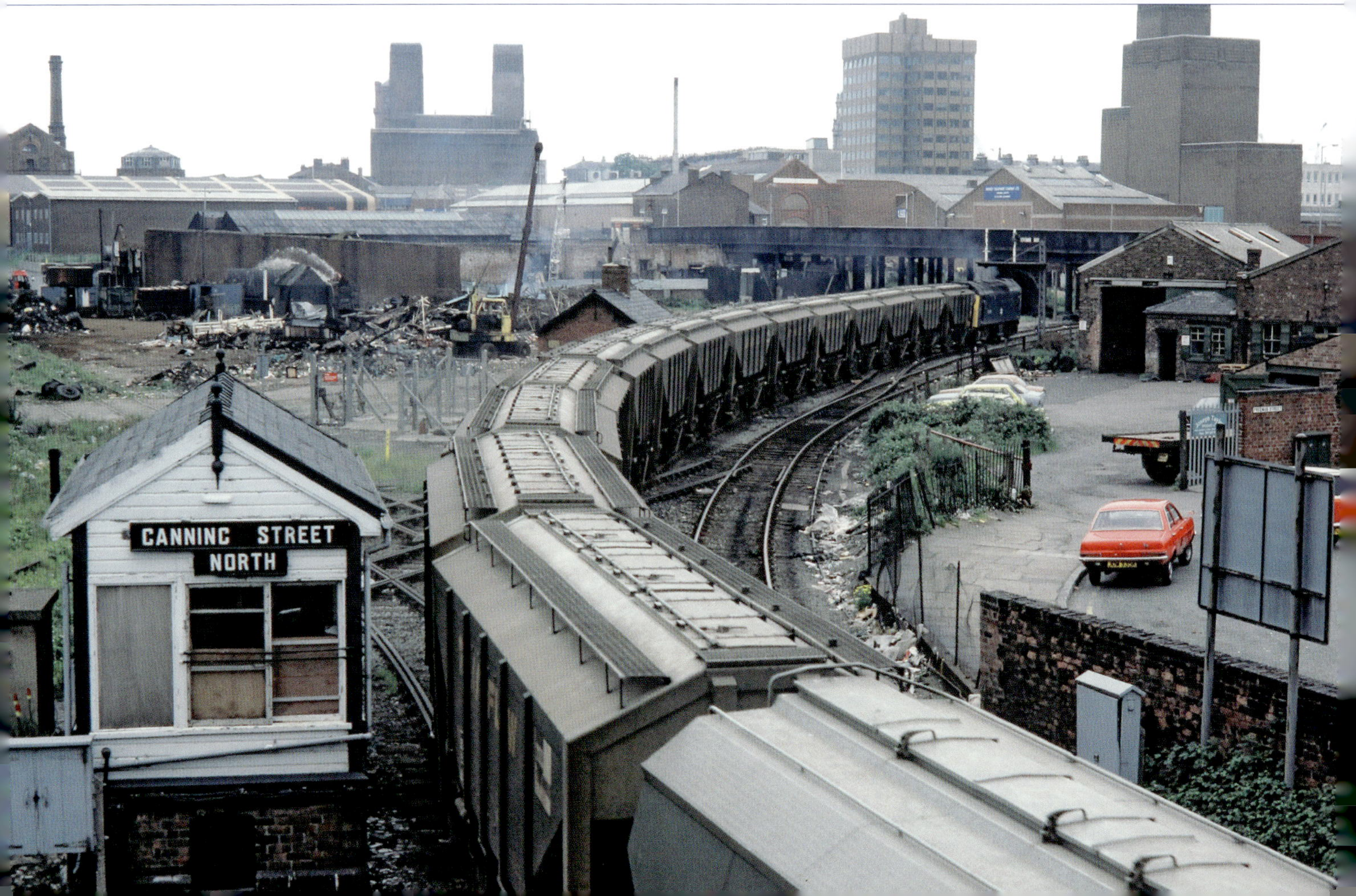
CANNING STREET
NORTH

Top left. Canning Street North signal box occupied a pivotal position on the Birkenhead dock system, though it is uncertain, given its dilapidated exterior but stubborn survival, when it finally ceased to be in charge of rail traffic movements. In an unofficial but authoritative list of signal boxes in operational service on the railway network going into year 2000 it kept its place, and as a physical structure there was enough of it remaining to justify its name in a published photo from 2009. But in practice there was no traffic to control after 1993 because of the de facto closure of the Docks access line following unauthorised removal of part of the track. The line to the docks from Rock Ferry through cuttings and a short tunnel was part of the Birkenhead Jt line dating from 1847, and the line emerged from the last cutting in front of the signal box at a complex set of junctions. On 15th June 1984 a train of Grainflow covered hopper wagons (with three larger Polybulk vehicles near the tail) passed along this section with class 25 no. 25296 in charge. Notice that a roadway passes under the first vehicles in the train with a crossing barrier pointing at the sky, suggesting that warning lights alone protected road traffic. The viewpoint was from the footbridge (demolished in 2019, perhaps along with the signal box) with a vintage brick building alongside. *IMG 1379*

Bottom left. The camera followed the train round the curve as the BJt line exits the Dock system. Although the train and signal box conceal much of the original track-work there is enough to show that at one time the line divided with tracks proceeding ahead to a CLC goods depot and sidings. To the train's right there is a broken off section of track which can be traced back to pass beneath the train and the other side of the signal box to head north on the Four Bridges route. This line gave access to the Alfred, Wallasey and Morpeth (and Egerton) docks which flanked the Mersey on one side, with the Great Float East on the other. On the photo's skyline on the right, the handsome building with the stepped sides is the ventilation tower for the Mersey Railway tunnel, and the ventilation building with two separate towers directly ahead serves the Queensway road tunnel exit at Sidney Street. *IMG 1381*

Above. In June (13th?) 1966 I snatched this image of one of the standard-gauge Drewry Car Co. diesel-mechanical 0-4-0 shunters which worked in the docks. Drewry was exceptional among loco builders for designing the engines but (until its final years) leaving the building to sub-contractors. This was *WH Salthouse* built by Robert Stephenson and Hawthorns (works 2590 / 1959). I missed the front of the train but it was propelling a 13-ton mineral wagon no. E294970 (it and its wooden-sided predecessor were both marked XP from earlier employments). The Drewry had rear-end assistance from a tractor displaying the name 'WJ LEE' which might have referred to the owner, the driver or the tractor itself. All this (and my photography) was taking place under the watchful eye of an employee of Customs and Excise. The signage makes the location obvious. I was reunited with *WH Salthouse* on 29th January 1985 when it and two other Drewrys *Pegasus* (VF 2270 / 1949) and *Kathleen Nicholls* (Drewry 2724 / 1963) were parked under the shadow of three Cunard 'reefers' (*Servia, Carmania* and *Carinthia*, all having refrigerated space for transporting fruit). They were berthed in Vittoria Dock while, like the locos, they awaited further instructions. *HF-3-28-24*

Above. Along Corporation Road and beyond Vittoria Dock the road meets Duke Street, where on 13th September 1983 there were active road works in progress making problems for the railway. Class 47 diesel no. 47334 is crawling towards the road junction passing two mechanical diggers with debris from their work by the lineside. There are pedestrians on the scene and a figure in black wearing an official hat of some kind stands with arm outstretched as if to ensure that nothing impedes the train's progress. The loco's driver is looking down from his side window and the second man in the cab is apparently doing the same the other side. Only two wagons behind the loco are visible, the first a CGV Grainflow with a 4-axle Polybulk Grainflow behind. The other side of the loco is an unlikely area of grass, and distantly on a large newish grey building is a sign for 'British American Product Ltd'. But adjacent to the loco are street signs advising traffic to go right (east) for Liverpool and Duke Street Bridge. East for Duke Street Bridge? Incidentally, at each end of the Grainflow Polybulk wagons there was a warning sign about hazards in English, French and German, but that of course was before Brexit. *IMG 0524*

Top right. The integration of the Birkenhead docks railway with its economic and social environment was even more apparent in this image taken from the same location as the previous (IMG 0524), but putting the Royal Duke centre stage immediately before no. 47334 with its train entered the road crossing. The Royal Duke with the names of Corporation Road and Duke Street on the building was a notable landmark to the area, though like much else in the docks it is now unfortunately gone. (Have the setts in the foreground been dug up for sale or tarmacked over?) Class 03 diesel shunter no. 03162 was stood on a sharply-curving siding from the Cross-Dock route, though as I had no idea whether the grain flow wagons were for the terminal to be filled or one of the mills to be emptied, I could only guess whether the shunter's assistance would be needed. A Merseyside Transport Atlantean double-deck bus no. 1588 (GKA 13N) showing Route 1 New Brighton as its destination was crossing into the southward extension of Duke Street. The only false note in this photo (further revealing my lack of familiarity with Birkenhead dockland) is that the bus was heading south, not north as I assumed it should if aiming for New Brighton. *IMG 0525*

Bottom right. The Mersey Railway (MR) opened in 1886 between James Street station in Liverpool (thence to Liverpool Central Low Level in 1892) and Green Lane station in Birkenhead, with a branch in 1888 to Birkenhead Park station. Next station on that branch would eventually be Birkenhead North, shown here on 7th February 1987 with two of the class 508 EMUs at the platforms and a third for 'Depot' parked alongside. The MR initially relied on the inadequate condensing gear of its steam locomotives (like the now preserved *Cecil Raikes*) to control the smoke pollution in the tunnel, but in 1903 it was forced to electrify its lines employing a 4-rail system which extended to Park station. An earlier station near the site of Birkenhead North was opened by the Hoylake Railway as Birkenhead Docks, but was relocated and named by the Wirral Railway (WR) in 1926. Arriving from West Kirby the steam-using WR bridged the gap to Park where through passengers were obliged to change trains. At the Grouping in 1923 the WR was absorbed by the LMS which electrified the line in 1939 but with 3-rail, leaving the MR (excluded from the Grouping) to adapt its 4-rail to ensure compatibility. Not until 1955 did the MR convert fully to the 3-rail system. The class 508 EMUs were built in 1979-80 and were introduced onto Merseyrail's Wirral Line service in 1984. *IMG 1600*

Merseyside and West Cheshire Railways 1965-1990

Merseyside and West Cheshire Railways 1965-1990

Above. Barely a mile west of Birkenhead North on the Wirral Line is Bidston station which, like its near neighbour, owes its existence to the Hoylake Railway and was established in 1866. With only a small population in the immediate area the station has always been reliant on proximity to the docks for freight (formerly) and as a point of access to another line for passengers (now). Its survival remained uncertain until 1896 when the North Wales & Liverpool Railway adopted it as its northern terminus, building south to Hawarden Bridge on the Dee. The NWLR was a joint creation of the Manchester, Sheffield and Lincolnshire Railway and the Wrexham, Mold & Connah's Quay Railway which failed in 1897 leaving the MSLR (renaming itself the Great Central Railway) in possession. As well as its connection to the south, Bidston also connected by Bidston North Junction to Seacombe and New Brighton. Visited on the 7th February 1987 looking west towards the Wirral Line terminus at West Kirby a Pacer DMU no. 142 028 was arriving from Wrexham (the branch left) passing Bidston Dee Junction box. Bidston's final claim to fame in steam days were the iron ore trains from Bidston Dock to the John Summers steelworks at Shotton hauled by 9F 2-10-0s. They finished in 1967, the Dock has since been filled in, the East and North Junctions curve lifted, and Bidston Dee Junction box demolished. *IMG 1597*

Top right. When Woodside station closed in November 1967, Birkenhead Town station had already disappeared in 1945 so that Rock Ferry was literally next in line and became the de facto terminus for the remaining passenger services to Chester and Helsby. The importance of Rock Ferry dated back to 1891 when the Birkenhead terminus of Mersey Railway's underground crossing of the river in 1886 was pushed further south from Green Lane station to Rock Ferry. There it shared the station with the main line from Paddington, and thus allowed passengers from Chester and beyond who wished to complete their journey to Liverpool by rail could do so by changing to the MR. On the 1st March 1967 in the last week of through trains, the 12.10 departure from Paddington was about to leave Rock Ferry on the last leg of its journey. The driver of 'Black Five' 4-6-0 no. 45353 was looking back for the guard's green flag to proceed. The train on the left of the picture was a class 503 EMU heading for Birkenhead Central and Hamilton Square before arriving at the terminus at Liverpool Central Low Level. Just visible on the right-hand end of the distant gantry is a lower quadrant signal arm – the GWR/LNWR legacy still present. *HF-4-94-19*

Bottom right. The 09.10 departure from Paddington on 22nd October 1966 was brought into Rock Ferry station behind BR Standard class 3 2-6-0 no. 76052 on its way to Woodside. Because of the triangular configuration of Chester's railways, trains arriving from the south (or north) reversed and changed locos in Chester station. For northbound trains with only about 15 miles to run to Woodside the new loco might well be from Chester loco shed as in this case, though it was recently transferred from Saltley. A feature of Rock Ferry, and an indication of its status as a shared station of the Birkenhead Jt with the MR, was the glazed pedestrian walkway which spanned the six platform lines, and the three lift towers (one off-picture to the right which served the MR platforms). These would have handled the transfer of luggage and parcels from platform to platform for the convenience of porters but probably not (or no longer) passengers. Painted maroon and cream the colour has faded rather, and they would not have received another coat before the station was rebuilt by the early 1980s. The houses in Lees Avenue back onto the railway so whoever had put out a line of washing that day would have welcomed the end of steam, still two years away. *HF-S1344*

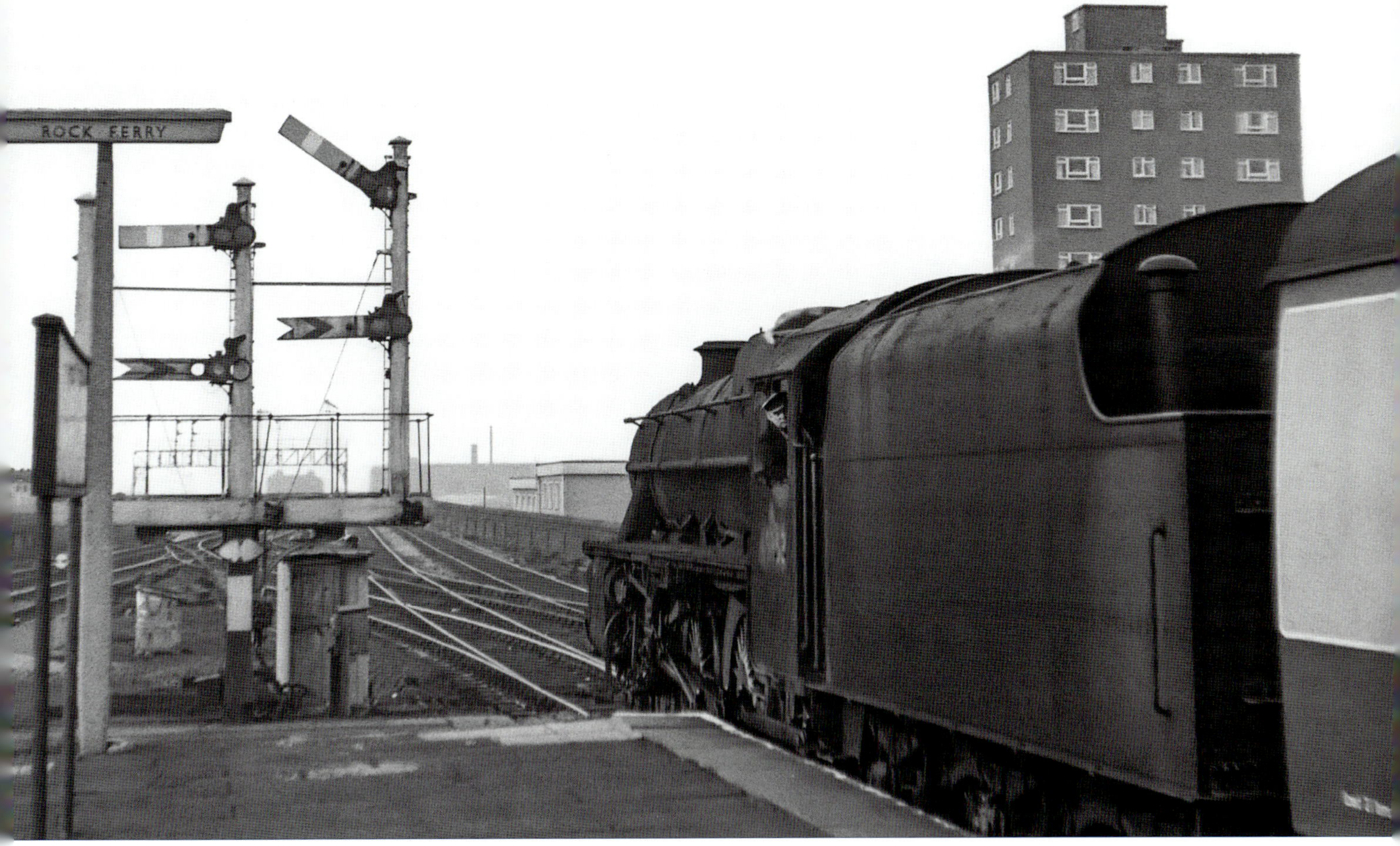
ROCK FERRY

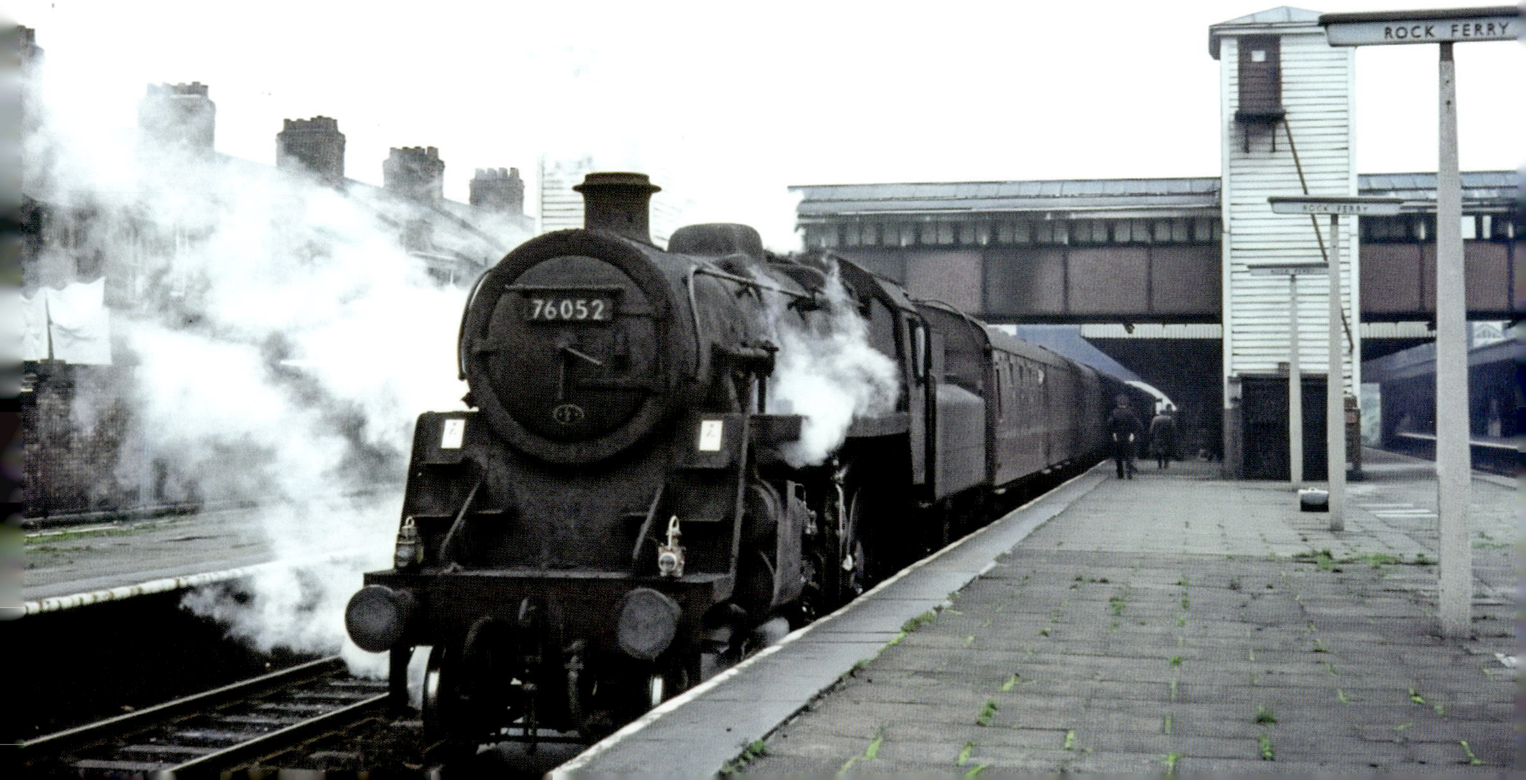
ROCK FERRY
76052

Above. Judging by the amount of coal piled high on its tender Stanier 'Black 5' 4-6-0 no 45201 is just starting its day's work. One of Speke Junction's allocation, its train is a collection of battered mineral wagons the first few of which for some reason looked empty as it passed but the remainder were well loaded with coal. This view of Rock Ferry station on 7th July 1967 is looking south towards a second covered bridge for passengers over the line and then the bridge under Bedford Road. The station entrance was (and still is) to the right of the picture, in 1967 overlooking the bay platforms of the former Mersey Railway. On the houses to the left there is an astonishing array of chimneys and pots, almost all of which have disappeared (though the houses are doing well), a reminder like the loco of our overwhelming reliance then on coal for domestic comfort and energy. *HF-S1760*

Top right. Hooton station is barely 5½ miles south of Rock Ferry but in that stretch there are now six stations. This is commuter territory, with only Bebington, Spital and Bromborough stations dating (like Hooton and Rock Ferry) from the 1840s. Two others came into existence following electrification south to Hooton in 1985. Only Port Sunlight had a major link to industry, from the 1880s with Lever Bros. works (now closed). But this line carried such a volume of traffic, much of it originating with or destined for Birkenhead Docks, that it was quadrupled in 1900-08 (from two miles south of Hooton) until being reduced to double in the 1970s. The photo shows Standard 9F 2-10-0 no. 92154 heading south beyond Hooton station with what looks like a train of metal castings, on 2nd January 1967. Distantly visible are Hooton South signal box which was replaced in 1985, and to the right a very imposing Water Pumping station of Victorian red brick. Between the loaded coal trucks and the brake van (left) but hidden between the two electricity pylons was the track-bed of the branch line from Hooton heading west across Wirral to reach Neston and Parkgate in 1866 before turning north to arrive at West Kirby in 1886. This line (which closed in 1956) was also part of the BJt's network (like of course the main line through Hooton station). *HF-4-43-8*

Bottom right. The heyday of Hooton station lasted a half-century, between the quadrupling of the line and the closure of the West Kirby branch in 1956. During that period the station staff managed three island platforms, and an entrance building which housed the offices and also a bay platform (platform 1). This was for the branch line shuttle service to Helsby, with in 1922 no less than 20 weekday departures. Platforms 2 and 3 were the up and down fast lines, 4 and 5 up and down slow. The West Kirby service (15 departures on weekdays) was handled by platforms 6 and 7 (the latter the near face on the nearest island. The platforms were accessed by an overhead walkway and steps (similar to the arrangement at Rock Ferry but without the lifts). The outstanding features of the station are the decorative brickwork on the outside wall and the square pyramid on the roof, both still surviving. The roof pyramid is just visible in this view from 5th March 1966, with Standard 9F 2-10-0 no. 92105 paused alongside platform 5 displaying its flangeless wheel(s) on the central driving axle. In the right-hand foreground are the remains of platform 7 where the track has been lifted. The water column and brazier are still there but the bag has been removed, whereas platform 5 adjacent to the loco has the full kit. Standing at platform 3 is a class 101 DMU on a Chester to Rock Ferry local service. *HF-2-26-32*

HOOTON

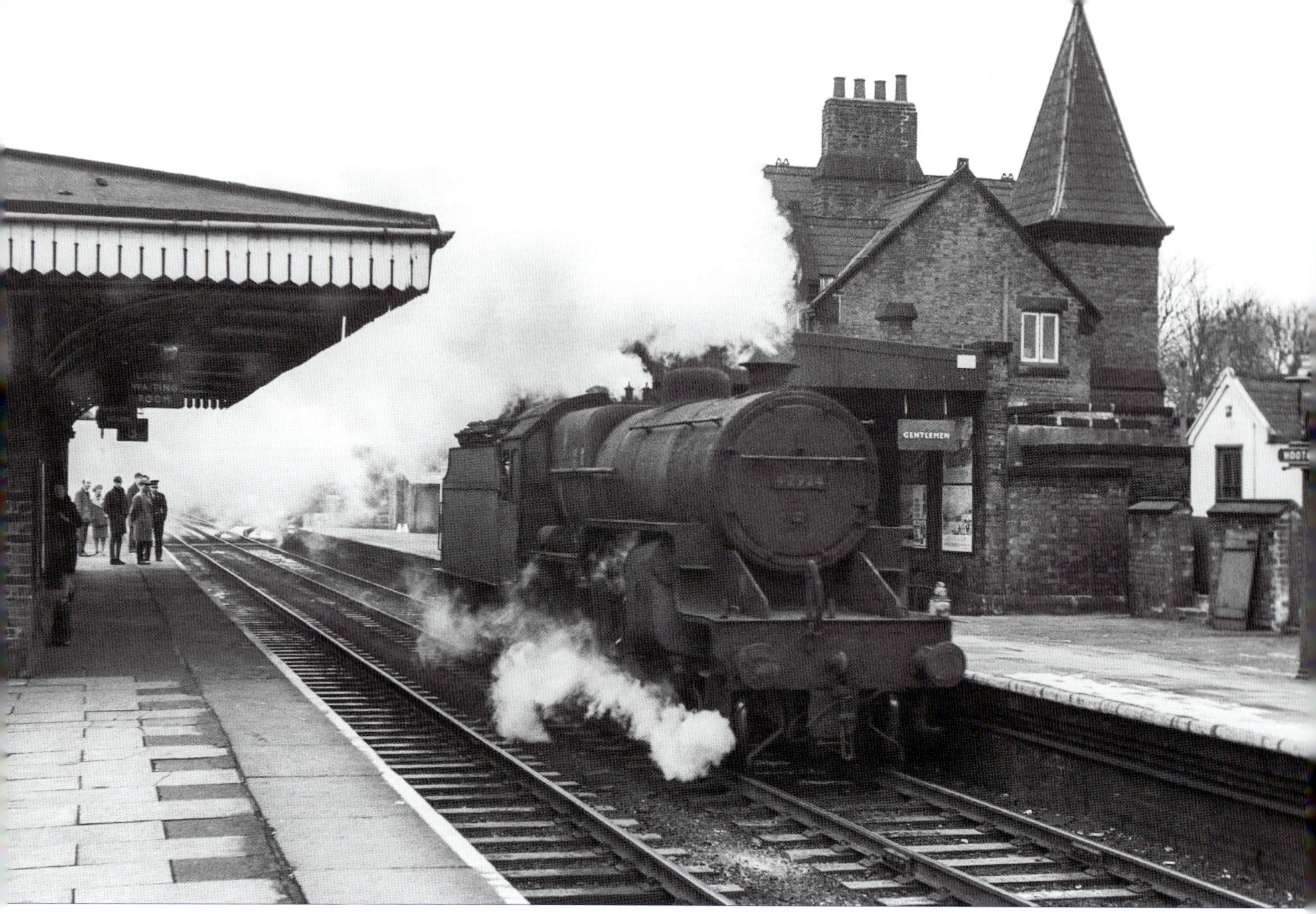
GENTLEMEN

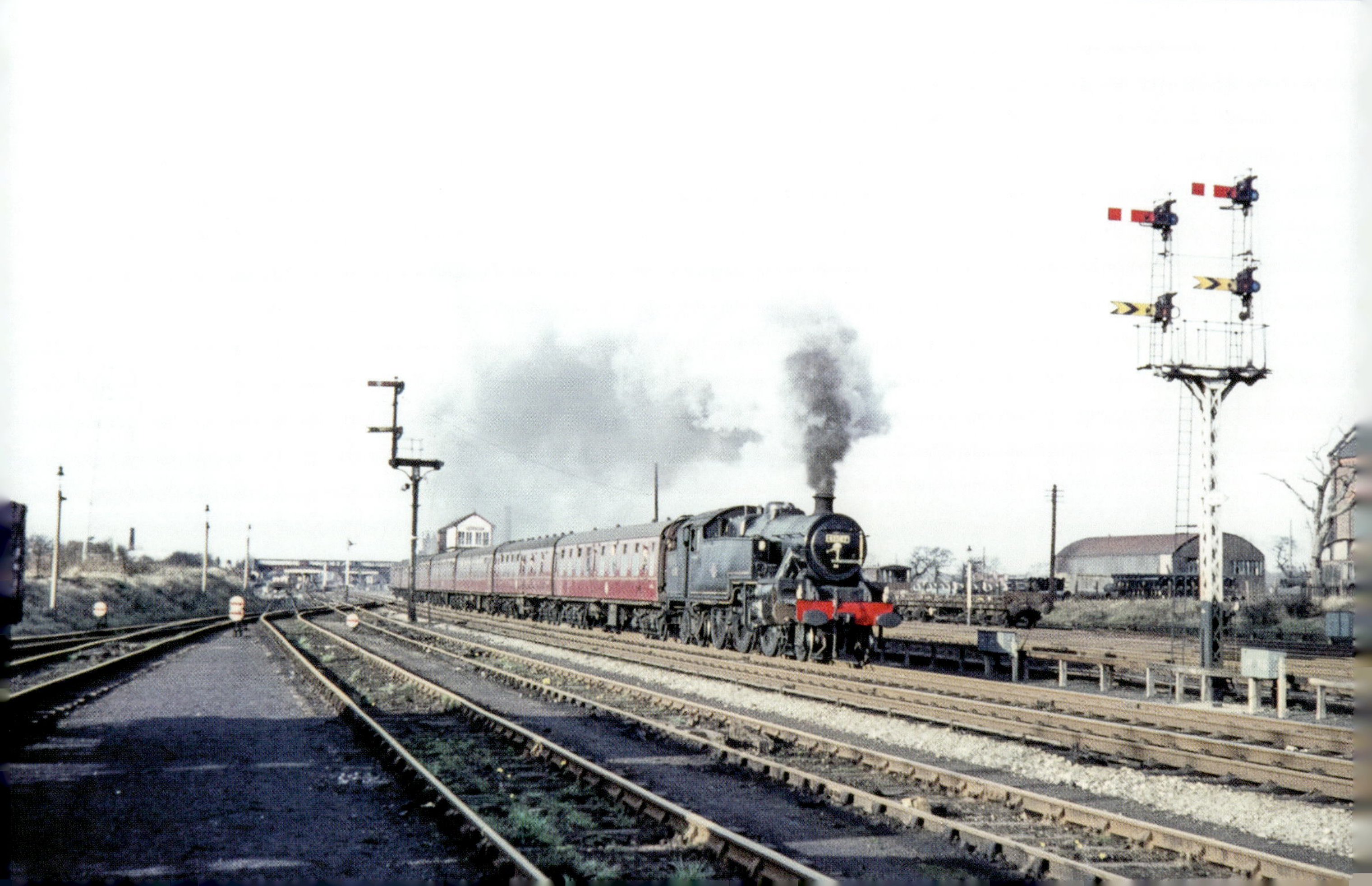

Top left. This image from 15th January 1966 shows one of the last remaining 'Crab' locos no. 42924 running southbound light engine alongside platform 2 and under the curious pyramid on the station roof. Six weeks later the loco was withdrawn from service. On platform 3 a group of passengers is waiting for a train (perhaps savouring the mix of steam and coal smoke for old times' sake before their DMU's arrival), but the figure closest to the platform edge ensuring the group's safety is a porter. It's hard to see but he wears the traditional British Railways hat that was replaced by a new style for BR later in the year. The 'Crab' was an LMS 2-6-0 of which about 245 were built during 1926-32 either at Crewe (as in this case in 1930) or at Horwich. It probably earned its nickname from its individual appearance, especially the rakish angle of the cylinder block, but with a power rating of 4 or 5MT it proved its value. The design began with Hughes, CME of the LMS until 1925, who was formerly with the L&Y (hence the loco was alternatively nicknamed the 'Horwich Mogul'). Fowler succeeded Hughes and sought to modify the loco along Midland Railway lines, not least by equipping it with an ill-matching tender. Stanier arrived at the LMS in 1932 from the GWR bringing with him a decided belief in the merits of the tapered boiler (so no more Crabs were built). *HF-2-4-41*

Bottom left. Sunday 5th March 1967 was the last day of through services between Paddington and Birkenhead, and the week-end saw a number of special trains arriving at Woodside to commemorate the final day of the connection. The glamour was provided by Castle class 4-6-0 no. 7029 *Clun Castle* hauling a Stephenson Locomotive Society special from Birmingham which, if indeed it joined the train at Birmingham, would have taken the Chester Curve to by-pass Chester station en route to Woodside. Built in 1950 by BR to the GWR design and withdrawn in 1965, it has not been continuously active in preservation but it visited Liverpool Lime Street in 2022. It made a fine sight sweeping through Hooton but Castle class locos were far from common at Woodside, so it is a more authentic commemoration to display this image of Stanier 2-6-4T no. 42587 leaving Hooton with the 15.25 service from Birkenhead to Paddington. The loco was cleaned for the occasion and seemed to be in fine fettle as it had been the previous day. As usual the train reversed in Chester station after detaching no. 42587 which then unfortunately failed, leaving 'Black 5' 4-6-0 no. 44690 to complete the last items on the diagram. *HF-S1454*

Above. Three years before building the Hooton branch, the BJt made a very much better investment in 1863 by opening another branch line which ran eastwards for 8½ miles from just south of Hooton station to Helsby. Intermediate stations included Little Sutton, Ellesmere Port, and Ince & Elton, with Stanlow & Thornton added in 1940. This line immediately gave more direct access for the LNW to its heartland, but also anticipated the opening of the Manchester Ship Canal in 1894 and the growth of industry along the route in the twentieth century. The most significant development for the railway was the Shell oil refinery at Stanlow which generated a huge amount of traffic, as this photo from 31st March 1966 suggests. Stanier class 8F 2-8-0 no. 48457 has taken the branch towards Helsby but the train of (empty) tanks indicates Stanlow is its destination. It has just passed Brush type 4 diesel no. D1860 (new from Crewe in August 1965) which is completely hidden from view behind no. 48457's train but has obviously just departed from Stanlow. It is about to join the main Birkenhead Jt line at Hooton where yet another trainload of tanks is arriving from the south behind an unidentified Standard class 9F 2-10-0. Completing the tableau in front of the 9F is what appears to be a rail maintenance vehicle. *HF-S312*

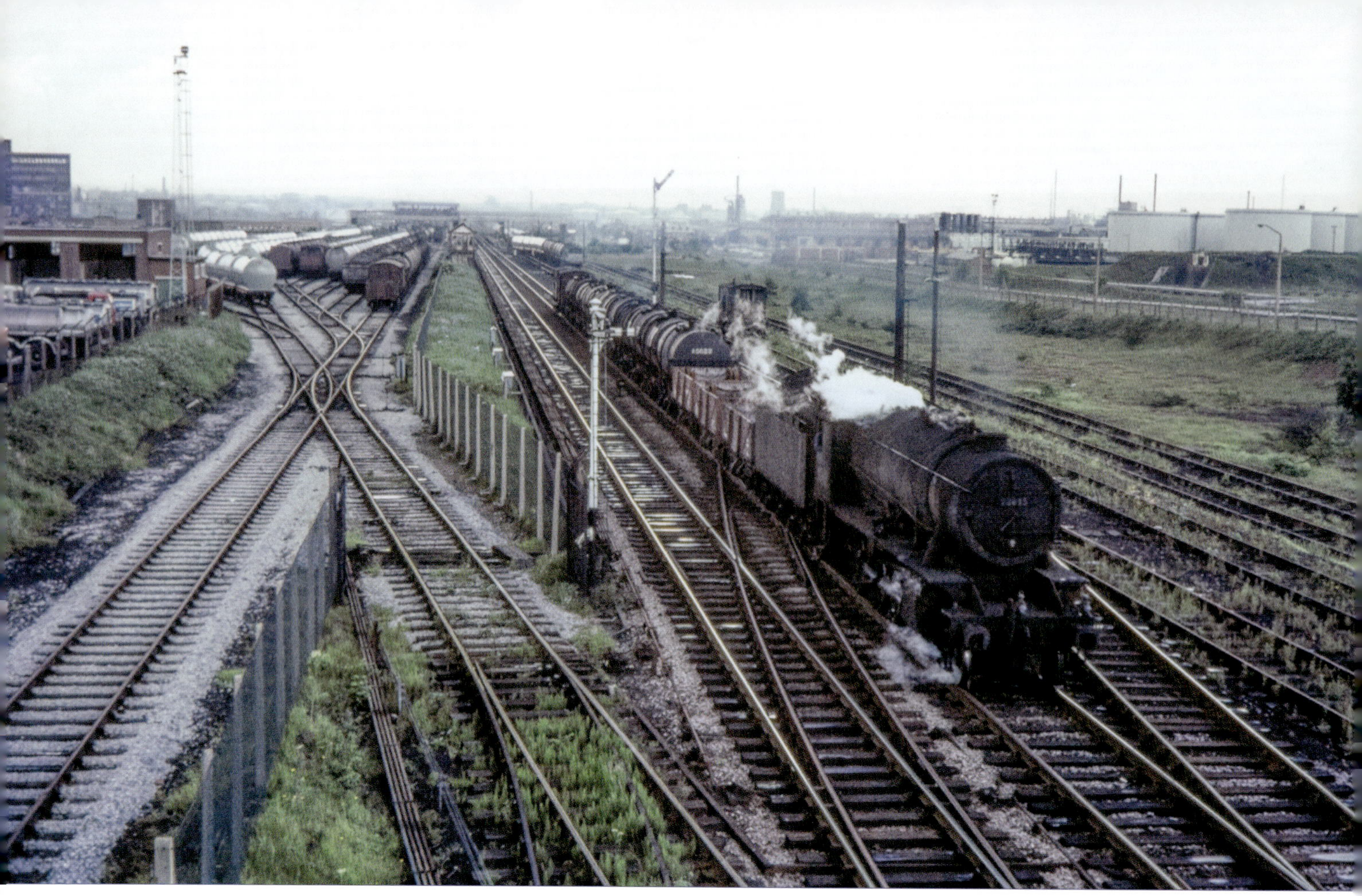

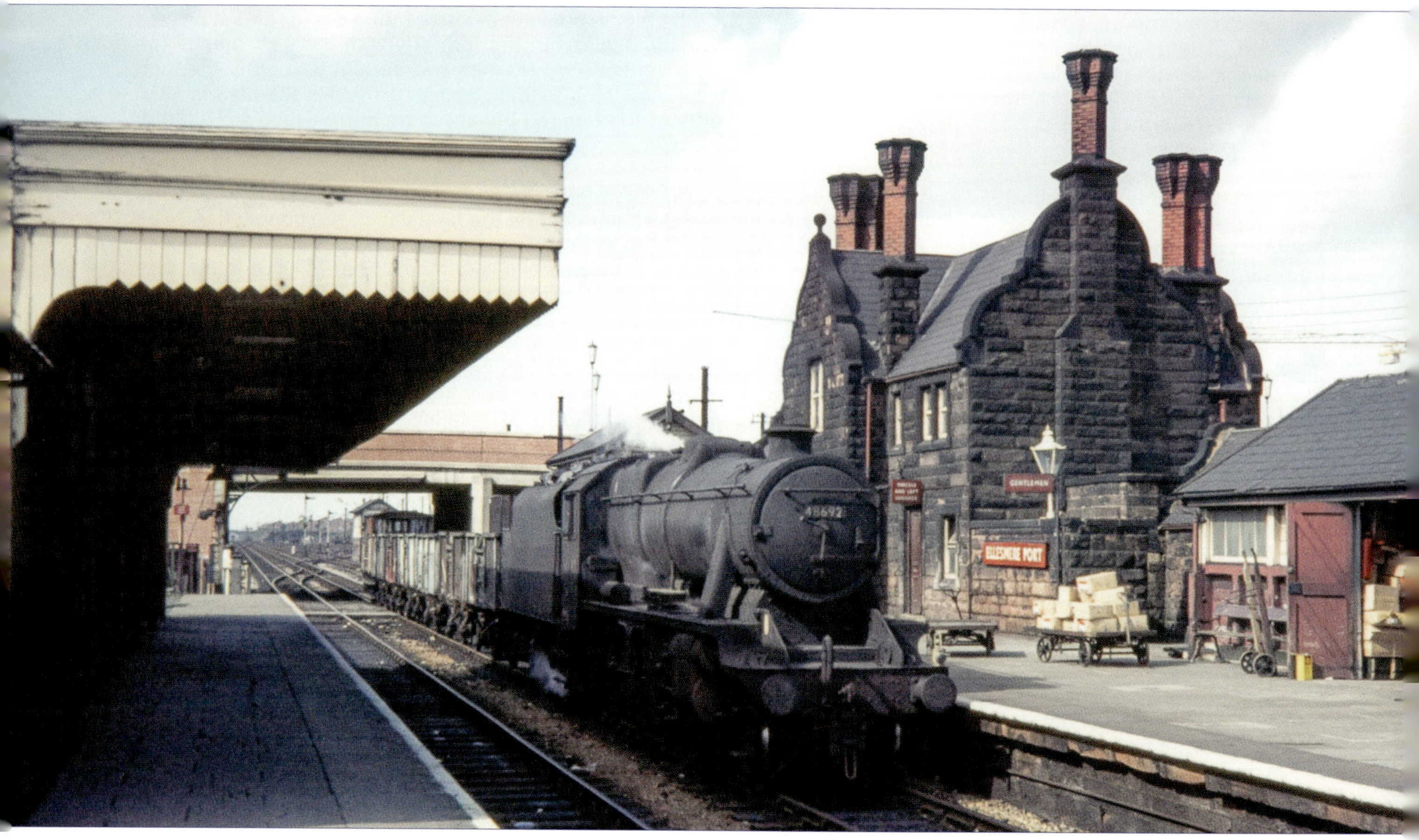
GENTLEMEN
ELLESMERE PORT

3 CHESTER

Top left. From modest beginnings in the 1920s, Shell Oil's refinery at Stanlow grew to become the UK's second largest in terms of refining capacity before it was sold to Essar in 2011. Stanlow's dependence on British Rail for distribution of its output was crucial up to the 1960s but eventually fell victim to the steady growth of the pipeline network after UKOP was created in 1969. Stanlow & Thornton station on the Hooton to Helsby line was opened in 1940 essentially to serve the needs of the workforce, with railway sidings taking their share of the seven square kilometres occupied by the industrial site. On a gloomy 18th June 1966 WD class 2-8-0 no. 90645 was heading eastward with a rather modest consignment mostly of tanks. At the far end of the lineside grass strip is the familiar gable outline of a signal box, though since it was new in 1940 it is rather surprising that Stanlow & Thornton did not get the LMS 'ARP' reinforced concrete roof protection of the also new (1940) box at Runcorn. Directly in front of the camera is the access line to the 14 reception sidings which are then squeezed into 5 lines to pass through the filling racks – the shed for these is just visible on the sky-line – before expanding again to form 13 departure sidings. *HF-S637*

Bottom left. On 31st March 1966 Stanier class 8F 2-8-0 no. 48692 was seen passing through Ellesmere Port station eastbound towards Helsby. The station looks busy with a large parcels and left luggage department with three trolleys and a barrow. I picture Joe Mercer, legend and great footballer, native of the town, as a boy (as he told it on radio) in the 1920s on the platform waiting to grab a bundle of the Saturday evening papers arriving from Liverpool with the football scores and a copper to be earned selling them around the pubs. But what really catches the eye is the wonderful architecture of the blackened red sandstone block building itself with its Dutch-inspired gable ends and lofty red-brick chimneys and elaborate pots. (Little Sutton station about two miles west also dates from 1863 but is smaller and less ornate.) Following the access of the Ellesmere/Shropshire Union Canal to the Mersey, the Ship Canal contributed hugely to Ellesmere Port's industrial growth, giving rise to an extensive rail infrastructure to the north of the town and sidings to the west alongside the main line. At one time five signal boxes were needed to control the traffic, one of which is unfortunately hidden just above the cab of the loco. Another is (almost) visible by the West End sidings, beyond the ugly A5032 Station Road bridge which replaced a station level crossing (but not until the 1950s). *HF-S310*

Above. At the west end of Chester station on the 14th May 1966 Stanier 'Black 5' 4-6-0 no. 45070 departs with the 08.55 service from Birkenhead Woodside to Paddington. The train will have arrived on the tracks passing signal box no. 4 ten or fifteen minutes earlier, time for the incoming loco to be detached and no. 45070 to take the train on to the south. Because of the restricted length of platforms at Woodside, northbound trains in earlier times would detach any surplus vehicles at Chester to be recovered on the return working. The triangle of lines is completed by the Chester Curve, a stretch of line which is unfortunately hidden in the photo but passes behind the water tower on the left to link with the route to Birkenhead. The Curve is rarely traversed except by special trains or as an alternative way of reversing. In the left foreground of the photo a Standard loco is backing down onto one or more vehicles in a bay. The building on the right had been the GWR's loco shed which even after 1948 catered primarily for GW locos, but since about 1960 dealt with the numerous DMUs employed on local services. The name of Chester station (ignoring Chester Northgate, the CLC interloper which arrived in 1875 and closed in 1969) was 'Chester' for the GJR in 1840, became 'Chester Joint' (about 1860 for the BJt [GWR – LNWR]), 'Chester General' sometime in the late C19th, and 'Chester' again in the late C20th. *HF-2-75-6*

Above. Lurking beneath Hoole Road bridge, vantage point for the previous photo, was 'Jinty' 3F 0-6-0T no. 47389. It was buffered up to carriages in one of Chester's west end bays, on 5th March 1966, awaiting authority for its next move. There's no sign of the fireman, who may have gone round to couple on to (or detach from) the stock. The water column stands ready if required but the ground at the platform end is dry so it's not been in use recently. The two braziers are unlikely to be needed again until perhaps the winter of 1967-8. By then no. 47389 would be just a memory, withdrawn from service three months after this photo. I confess I rarely enjoyed trying to photograph in Chester station itself. The external facade is very fine but the interior was a let-down, with no pattern to individual sections of roofing with obtrusive brick supports creating a rather gloomy atmosphere. The great stations of the north-west were termini with huge overarching roofs. Chester handled traffic leaving the city on six major routes in all directions and the station was substantially rebuilt with the provision of an island platform in 1890. It would not have helped in reaching agreement on this enlargement that the relationship of the two companies of the BJt was fraught with disputes. *HF-2-26-37*

Top right. In October 1978 class 40 1Co-Co1 diesel-electric no. 40106, the last of the class still clad in fading green, went into Crewe works to receive the regulation blue livery and emerged with a pristine coat of green complete with the old 'ferret and dartboard' BR insignia rather than the 'double arrow' which had been adopted in the 1960s. In its new retro guise the loco was immediately in demand for special trains but also earned its keep with freight as well as passenger duties, and was at Chester on the 4th September 1982 returning to Manchester Victoria with the 15.17 departure from Holyhead having worked the outward leg that morning at 10.45. By now work-stained, the loco was nonetheless displaying the accolade of 'royal train' headcode discs. Behind the elevated walkway and unfortunately almost completely hidden by it is a curiosity, Chester No. 3 signal box. This was installed at the turn of the C19th and originally controlled the crossovers which enabled the long up and down main lines to accommodate more than one train. (In the photo just the base of the box can be seen, with a supporting bracket on the wall.) Needless to say it did not survive Chester's resignalling in 1984, but the station's appearance has benefited somewhat from the Chester Renaissance Project in recent times. *IMG-0932*

Bottom right. On 20th March 1986 another class 40 but in a new guise as no. 97407 was pictured crossing the Shropshire Union Canal at Chester with a train of ballast hoppers from Penmaenmawr quarry along the North Wales coast. The loco was built in 1959 at Vulcan Foundry as D212, named *Aureol* in 1960, and renumbered but nameless as class 40 no. 40012 under the 1974 TOPS scheme. In January 1985 time was called on the few remaining active class 40s, but renumbered again as no. 97407 this loco was one of four that were withdrawn from service and taken into departmental (non revenue-earning) stock (the yellow rim to the wagons signifying 'departmental'). The reprieve for no. 97407 was short-lived with final withdrawal coming in April 1986, a month after this photo. In the foreground are the canal's Northgate Locks, a flight of three staircase locks designed by Thomas Telford in about 1790 and grade II listed. The larger expanse of water below the loco is part of the canal basin and junction leading to Crane Wharf on the River Dee. The elegant Georgian building beyond the railway (listed as Diocesan House, also grade II, and of the same period as the locks) has had various names and uses. The neighbouring red brick building is known as Telford's Warehouse (grade II) and is now an entertainment venue. *IMG-0834*

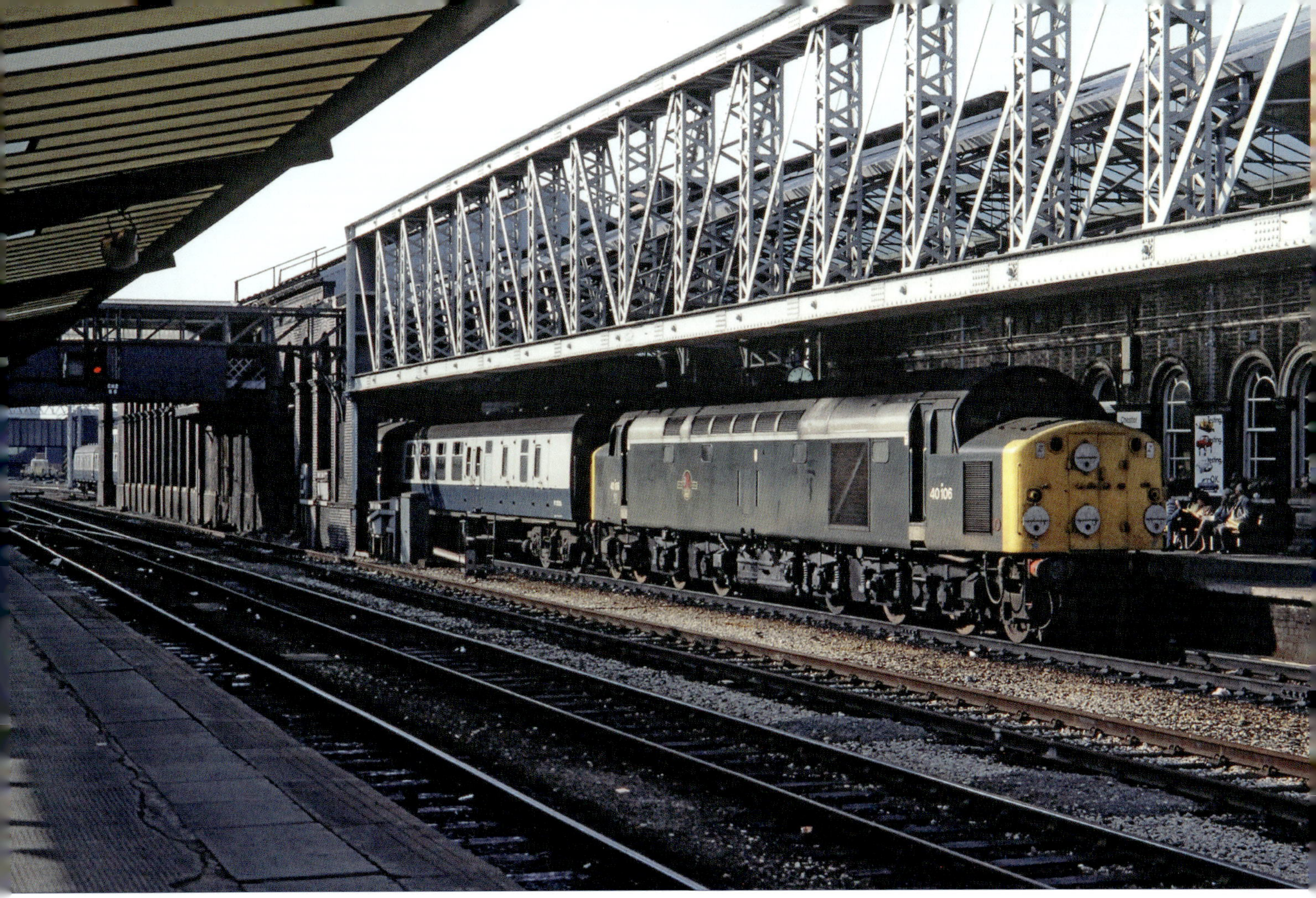
40 106

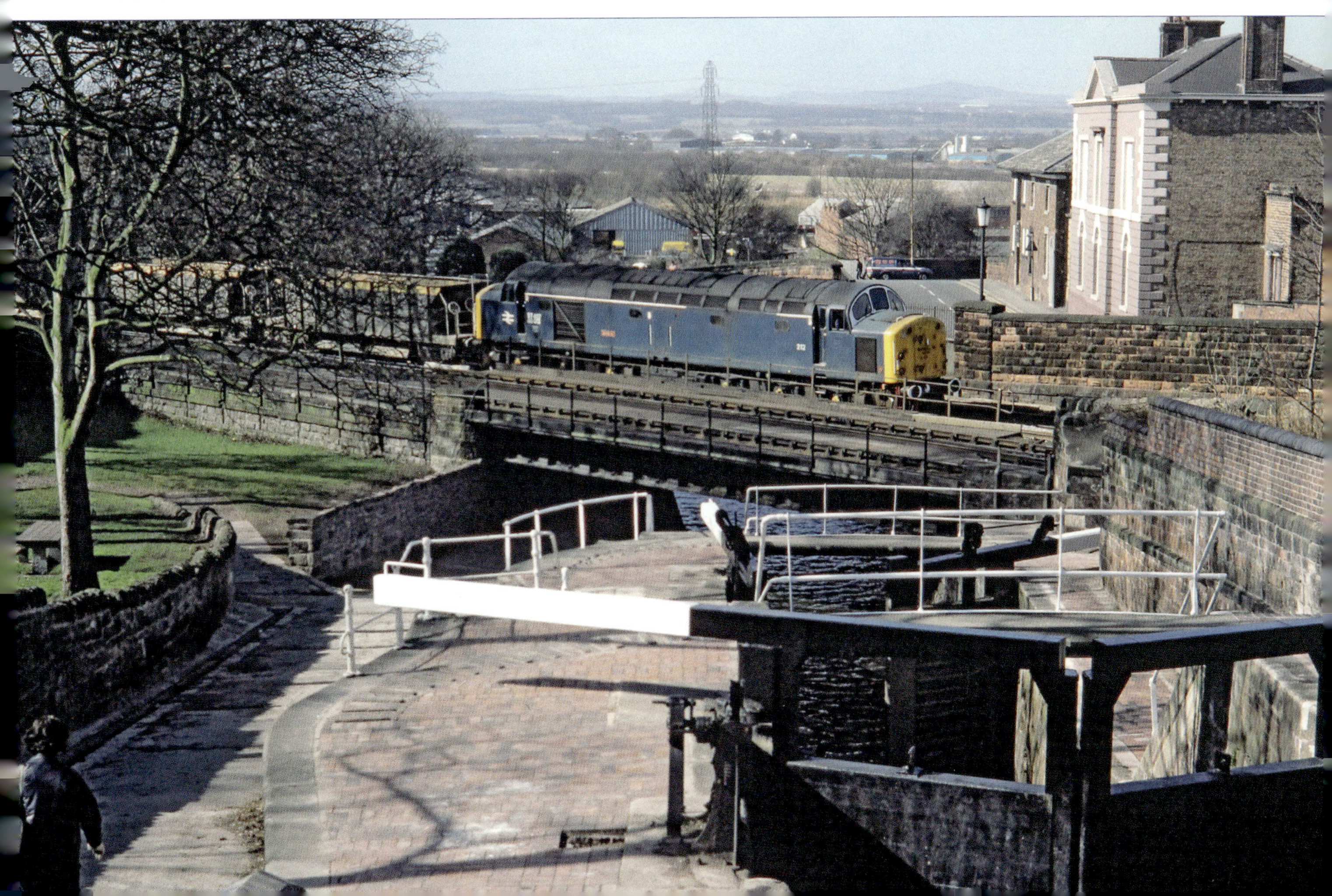

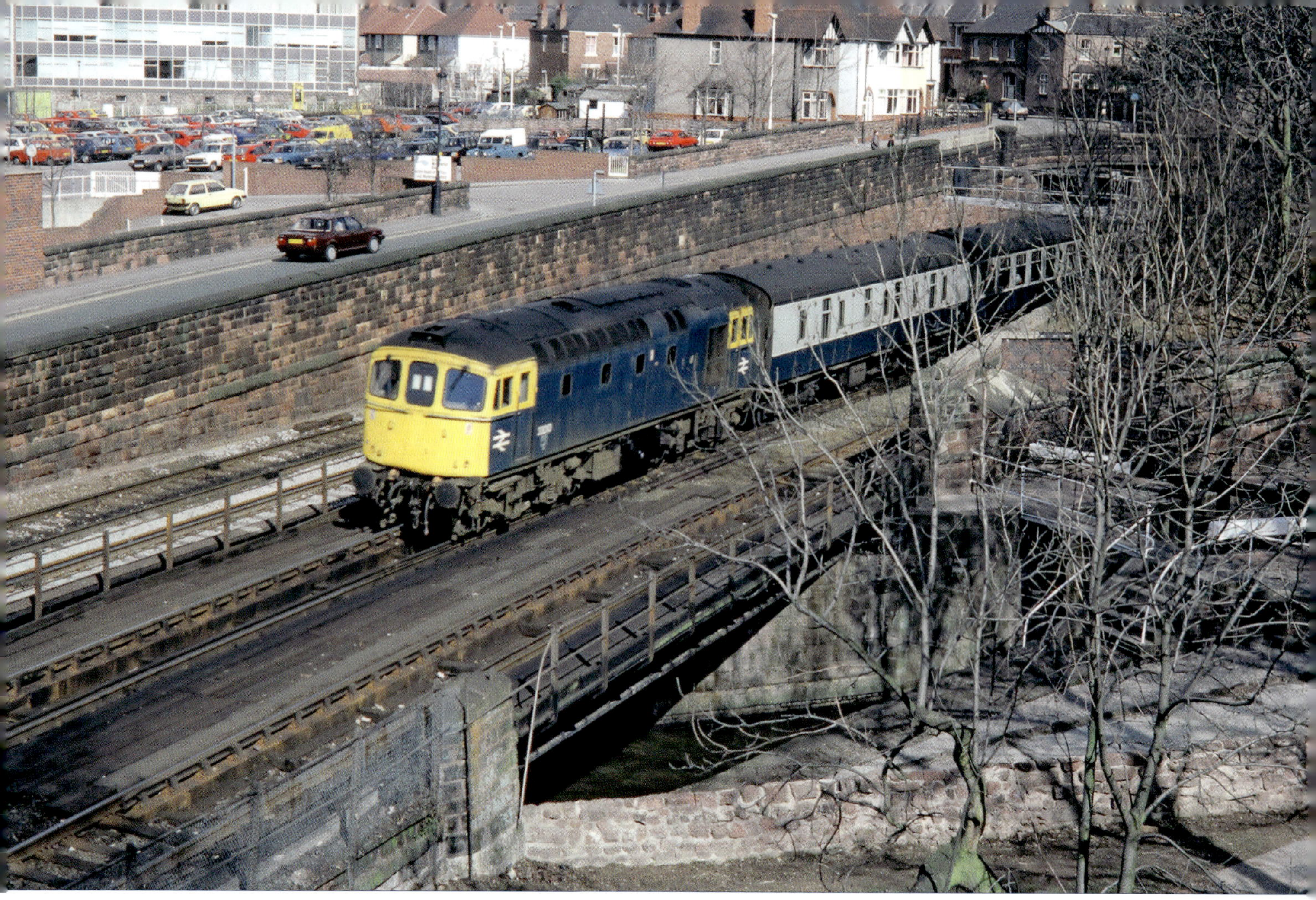

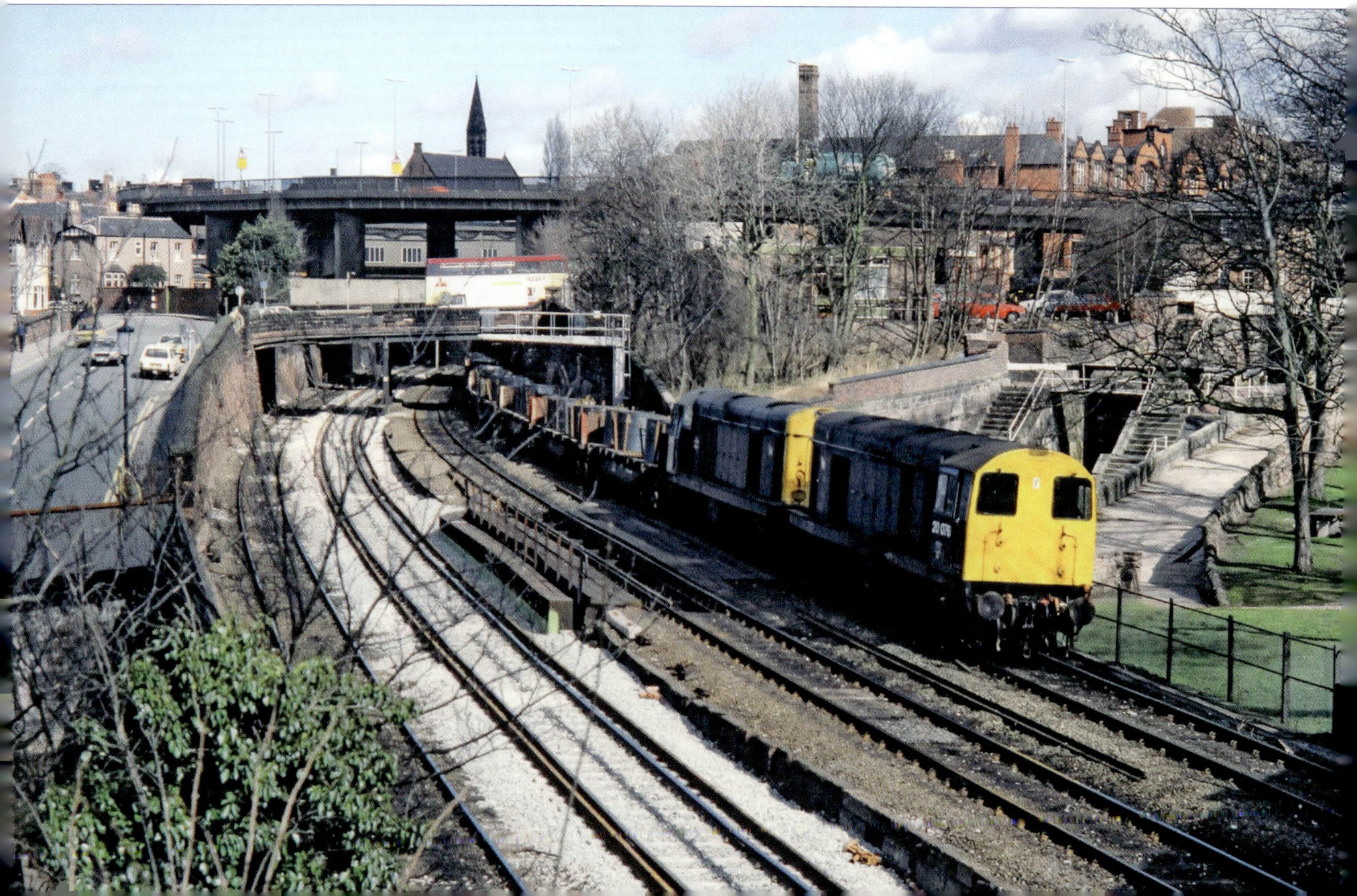

Top left. From about the early 1980s an unusual class of locomotive began to appear in the north-west, working services between Cardiff and Crewe. These were Type 3 class 33 Bo-Bo diesel electrics of which 98 were built mostly during 1960-1 for BR's Southern Region by the Birmingham Railway Carriage and Wagon Works Co. at Smethwick. They were Sulzer powered, with generators and traction motors by Crompton Parkinson. At first they were found mostly in the South-Eastern Division of the SR, on freight and secondary passenger work, but later they were also employed more commonly in the South West. As well as the cross-country Cardiff-Crewe route they also appeared in North Wales. On 15th January 1986 no. 33050 had worked the 11.15 departure from Crewe to Bangor, with class 47 no. 47457 as pilot on a double-headed train while running-in and immaculate fresh from works. In the photo no. 33010 was on its own in charge of the same service on 20th March 1986, and was seen from the city walls crossing the Shropshire Union Canal with Raymond Street beyond. Built in the 1960 batch as no. D6510, the loco was withdrawn in 1988, rather earlier than the majority of the class, many of which were still active in the 1990s. *IMG-0835*

Bottom left. No doubt it is self-indulgent to include three images of the same location on the same day, but the morning of 20th March 1986 was clear and bright, the railway traffic had plenty of interest, and the mix of canal and locks, bridges and tunnels, and roads at three levels, was hard to ignore. The vantage point this time was by Bonewaldesthorne's Tower, a 90° angle on the city walls which the Chester and Holyhead Railway negotiated by burrowing under the walls. The two locos on show here are English Electric Type 1 class 20 Bo-Bo diesel electrics, nos. 20076 and 20035. The 20s are celebrated because, amongst so many failures of types and classes ordered under the 1950s Modernisation Plan, they were a success. The first 20 were built in 1957-8, about 108 more in 1960-2, and a further 100 in 1966-7. All were powered by EE's 8SVT MkII engine which produced a slightly higher -pitched whistle than the sixteen cylinder version of class 40's16SVT MkII. Class 20 locos frequently worked in pairs, nose to nose, as here. No. 20076 was built in 1961 as D8076 and withdrawn in 1988. No. 20035 was built in 1959, withdrawn in 1991, but then remarkably was sold to work in France relabelled as the SNCF's Departmental CFD No. 2001, based at Autun for freight duties around Dijon. The loco arrived back from France in 2005 to provide spare parts for other 20s in preservation. *IMG-0837*

Above. Although Stanier's 'Black 5' 4-6-0s have always been highly regarded, the LMS sought to test further improvements to the design in the railway's final years before nationalisation, while Ivatt was CME. One such variation was to fit a double blast pipe and chimney, in a few cases combined with other experiments, but only nos. 44765-6 received just the exhaust modifications when turned out at Crewe in the last month of the LMS, December 1947. On a fine 28th October 1966, looking south into a pinkish morning sky, no. 44766 is seen approaching Chester's city walls. It has passed Roodee signal box where it appears semaphore signals on the down lines have already been replaced on the up by colour lights. On the right hand side are the Roodee Gasworks which closed in 1970 but whose gasometers remained a colourful landmark for many years. The train itself is made up of insulated containers, probably with meat from Ireland (via Holyhead). The first and third containers are white and sat in wagons, the rest are pale blue with the black arrow door-to-door symbol (visible as the train passed) on Conflats – short wheelbase flat wagons for transport of containers – to which they are tethered by chains. Loco no. 44766 had for a short time been a Llandudno Jct engine, but when 6G shed closed in October 1966 it was reallocated to Chester. *HF-S1353*

Above. The view of Chester from the south-west across the Dee bridge (aka Roodee viaduct) is a classic railway landscape. The Roodee is the name given to the 65 acres of grassland between the Dee river and the city which constitutes Chester race course. The race track itself is just over a mile long and racing takes place on about 15 days a year. Quite what happens on the turf for the rest of the year is not clear, but in spite of the white-clad figures in the photo organised cricket does not seem to be one of them, certainly not on the 5th March 1966. That day Stanier 'Black 5' 4-6-0 no. 45043 was crossing the bridge tender-first with a guards van at each end with running round at Saltney a possibility. Since 1966 there have certainly been changes to this view, including removal of the two tracks on the west side of the bridge. The race course grandstand (1899-1900) close to the city was burned down in 1985 – rebuilt by 1988 – while the principal visual features of Chester's skyline – from the left the Town Hall, the Cathedral and the Guildhall – are gradually being obscured by new developments. *HF-S167*

Top right. On the morning of 14th May 1966 Westminster Road bridge gave a fine view of the eastern end of Chester station. 'Jinty' 0-6-0T no. 47389, previously encountered as station pilot in March, was now shunting the goods yard established by the LNWR in the 1880s, alongside Lightfoot Street. The train consists mostly of vans but with a few mineral wagons no doubt to be separated from the rest. The figure in the cab door doesn't look like footplate crew, his hat suggesting a guard. The signal arm controlling the loco's movement is upper quadrant, perhaps surprising given the obvious age of the wooden structure. Sharing centre stage is the magnificent Chester No. 2 signal box which was demolished in 1984. To the left of the box is a signal gantry which has been stripped of all its signal arms except one on the right which is 'off', either for access to coach sidings just visible to the left of the station, or to the station itself. To the right of No. 2 box are two more home signals with probably more behind the box. Why was No. 2 box built so high? Was it to give signalmen a better view of events beyond Westminster Road and Hoole Lane bridges, or because the 182-lever frame needed the additional space? *HF-S444*

Bottom right. In 1870 the LNWR built a new loco shed to the east of Chester station, in a spacious V-shaped area between the GJR 1840 line to Crewe and the BL&CJR 1850 line to Warrington. This view looking south-east on 14th May 1966 from Hoole Lane bridge gives a general view of the shed area, as Stanier 'Black 5' 4-6-0 no. 45373 approaches Chester on the GJR route from Crewe with a mixed train of vans and parcels stock. The subsidiary signal arm was off and may indicate that the train was cleared to enter one of the bay platforms, or more likely sidings to the south of the station. The new (1870) shed itself was an unremarkable 8-road structure, with the only rail access by the line in the foreground. The turntable is unfortunately concealed by steam from the loco. The main structure in the yard was a water tank bridging across two tracks. Next feature of the shed yard viewing right to left is the system employed for coaling using a long line of mineral wagons, an arrangement which merits a closer examination. Last is the dump in the most remote corner of the shed yard where two 'Black 5s' await removal. *HF-S432*

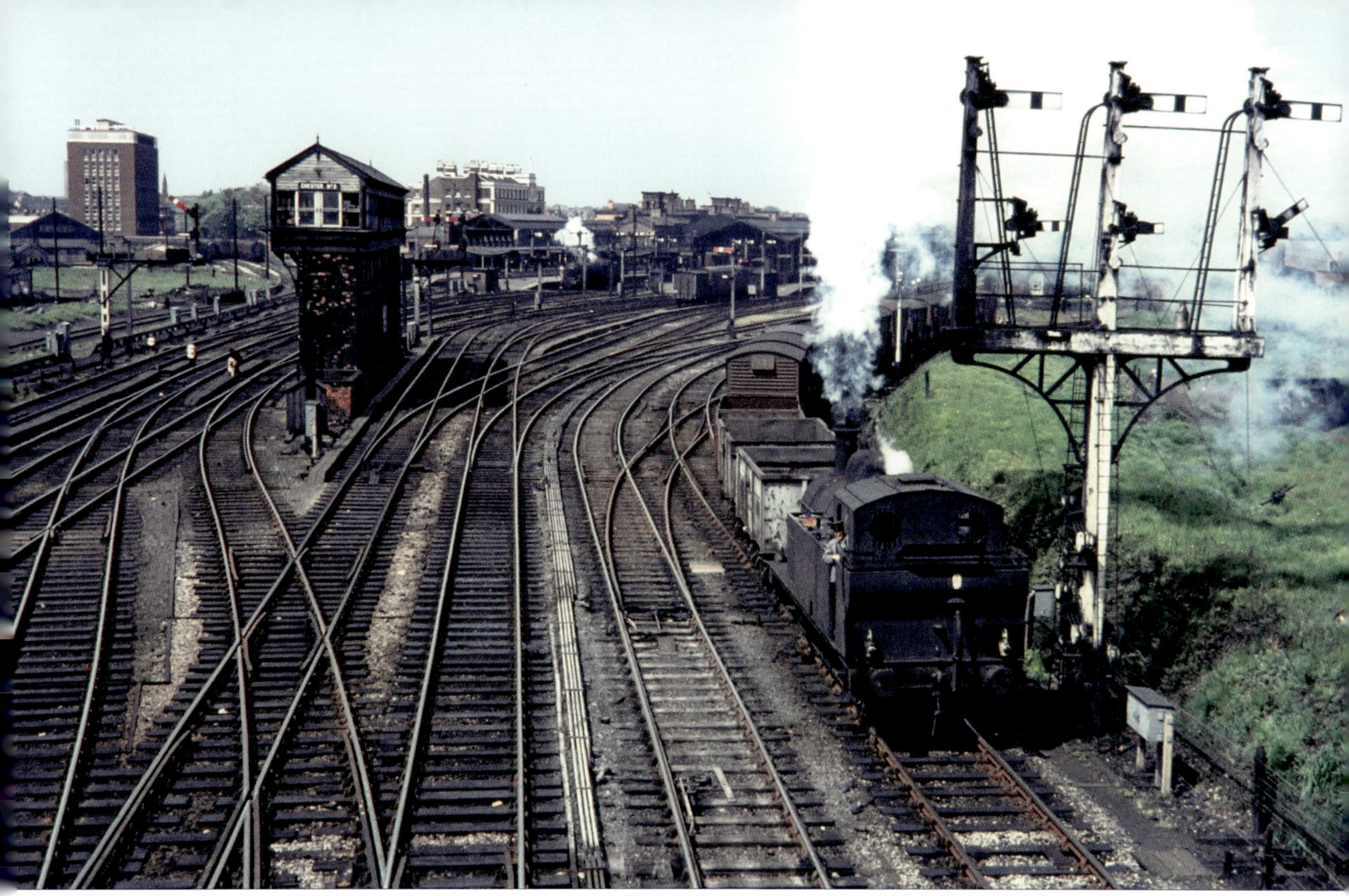
CHESTER No 2

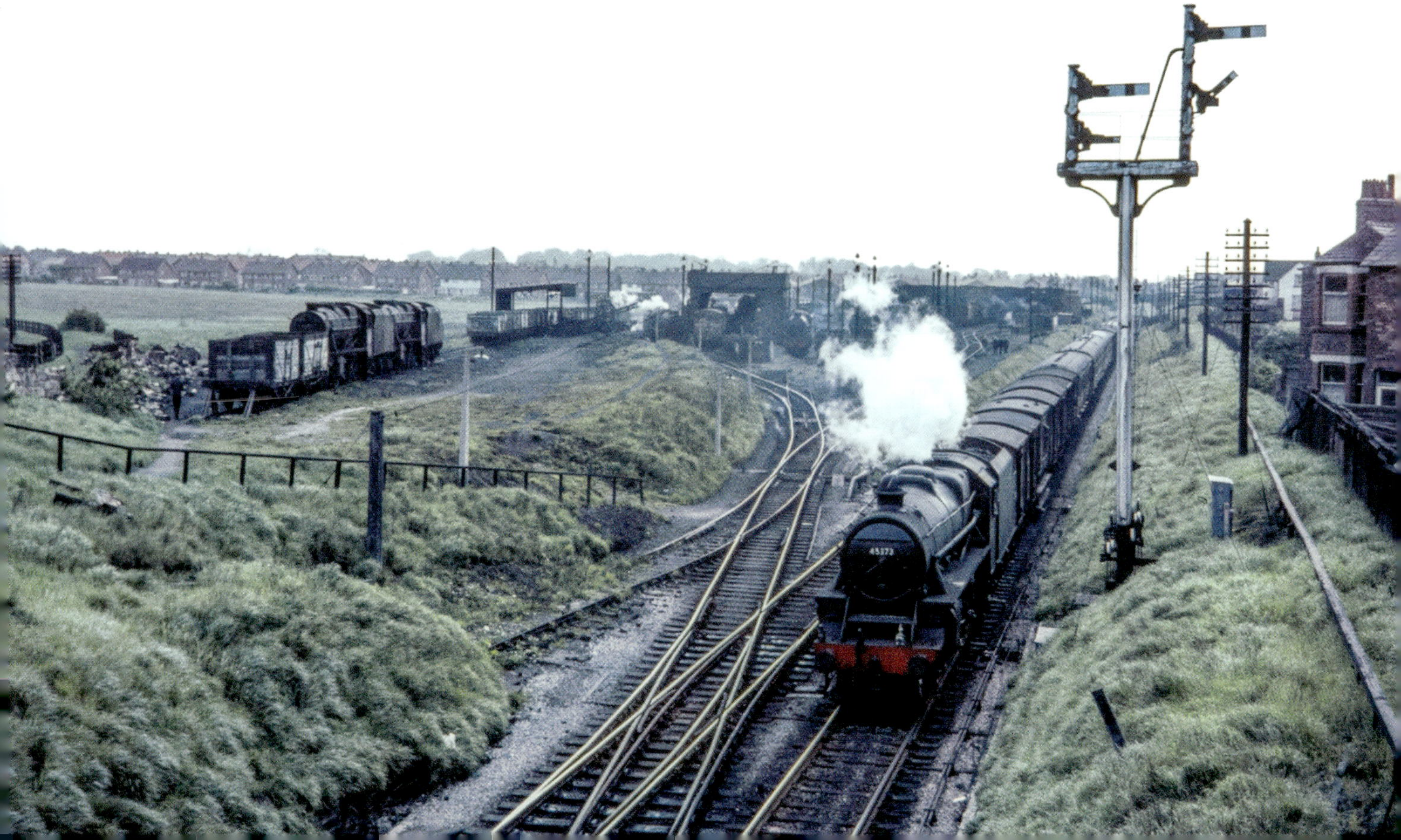

Above. A universal problem encountered by all steam railways using coal (or wood) is how to get tons of the fuel into a loco's tender or tank engine's bunker. A variety of ad hoc solutions was arrived at by British railway companies and by individual sheds, but especially in the 1930s and 1950s the device of choice became ferro-concrete towers in which coal was raised in wagons or skips and tipped into a hopper for eventual delivery to loco tenders waiting underneath. About a hundred of these towers were built, nearly half at LMS or LM region sheds including Aintree, Bank Hall, Birkenhead, and Crewe N and S. Standing about 70ft high it was obvious from any distance that Chester did not have one. But on my first visit to the shed on 5th March 1966 I was amazed to see Standard Britannia class 4-6-2 no. 70051 (formerly *Firth of Forth*) being coaled from a mineral wagon, the coal being shovelled directly from wagon onto a conveyor elevator (power source not known) by one or two lads, for tipping into the tender on an adjacent, slightly lower track. It looked amazingly primitive. *HF-S165*

Top right. Two months later (14th May 1966) the elevator for coaling was still in use. Standard 9F class 2-10-0 no. 92051 had travelled down the track tender-first to be coaled, in preparation for its next job which would probably have seen it leaving Chester on one or other of the Warrington and Crewe lines. A crude shelter gave some protection against the elements to the coal shovelers and two electric lights hung down, so this was not an emergency solution to the failure of a more sophisticated device. The conveyor elevator had wheels and to that extent was mobile, but in practice when each wagon was emptied of its coal it was the job of 'Jinty' Class 3F 0-6-0T no. 47507 to shunt the line a wagon's length. The shed code plate of no. 47507 is 6G – Llandudno Junction – where it was reported in October 1965 to be "maintained in sparkling condition" as station pilot. Since it was likely to be taken out of service within months (in fact in September 1966) it would hardly have been anyone's priority to replace the plate with '6A', especially as so many were by then being removed as 'souvenirs' or investments. *HF-S434*

Bottom right. The other eastbound route from Chester was the 1850 line of the Birkenhead, Lancashire and Cheshire Junction Railway which ran to Warrington. This company had absorbed the Chester and Birkenhead Railway (CBR) in 1847, changed its name to Birkenhead Railway in 1859, and was acquired by the LNWR and GWR for joint operation as the BJt in 1860. The 1850 line entered areas where the LNWR was becoming dominant, especially after 1869 and the Mersey Bridge opening, but the GWR had secured running powers towards Warrington and even Manchester through earlier acquisitions, hence the occasional appearance later of GWR locos in 'foreign' territory. There was nothing contentious about the appearance of Standard Britannia class 4-6-2 no. 70023 (formerly *Venus*) on 14th May 1966 with the 10.45 Manchester Victoria to Llandudno service. It is seen here passing the north flank of Chester shed (and a group of young locospotters). The BL&CJR had additional traffic that day, as after a serious accident the previous night the WCML was closed between Weaver Jct and Acton Grange Jct. Northbound trains (southbound vice versa) took the Runcorn line to Halton Jct. where they reversed on the LNWR's 1878 Halton Curve to Frodsham Jct. before proceeding to re-join the WCML at Acton Grange Jct. south of Warrington. After an interval of about 40 years with no regular passenger service, the Halton Curve had service between Liverpool Lime Street and Chester restored in 2019. *HF-S446*

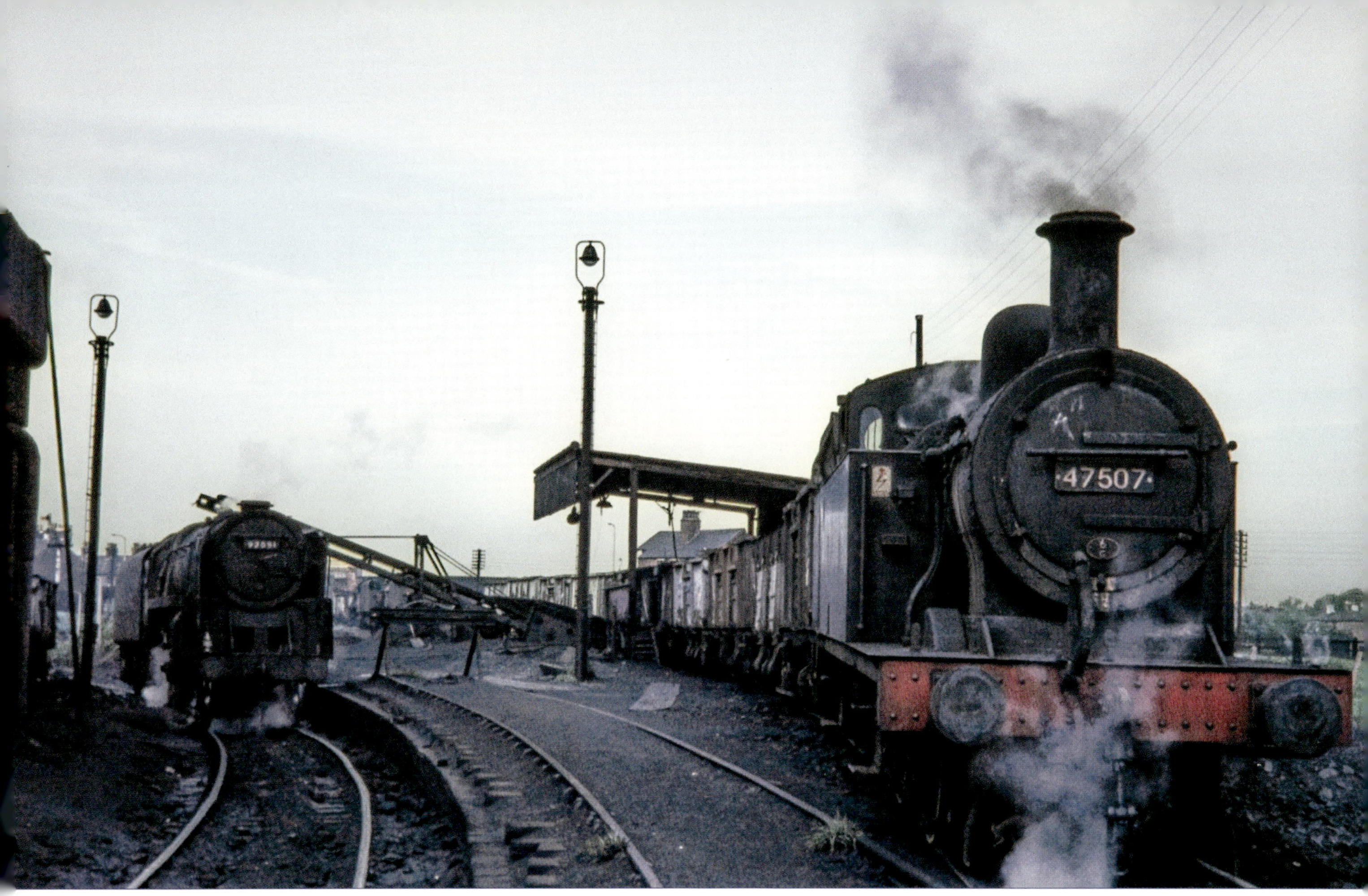
47507

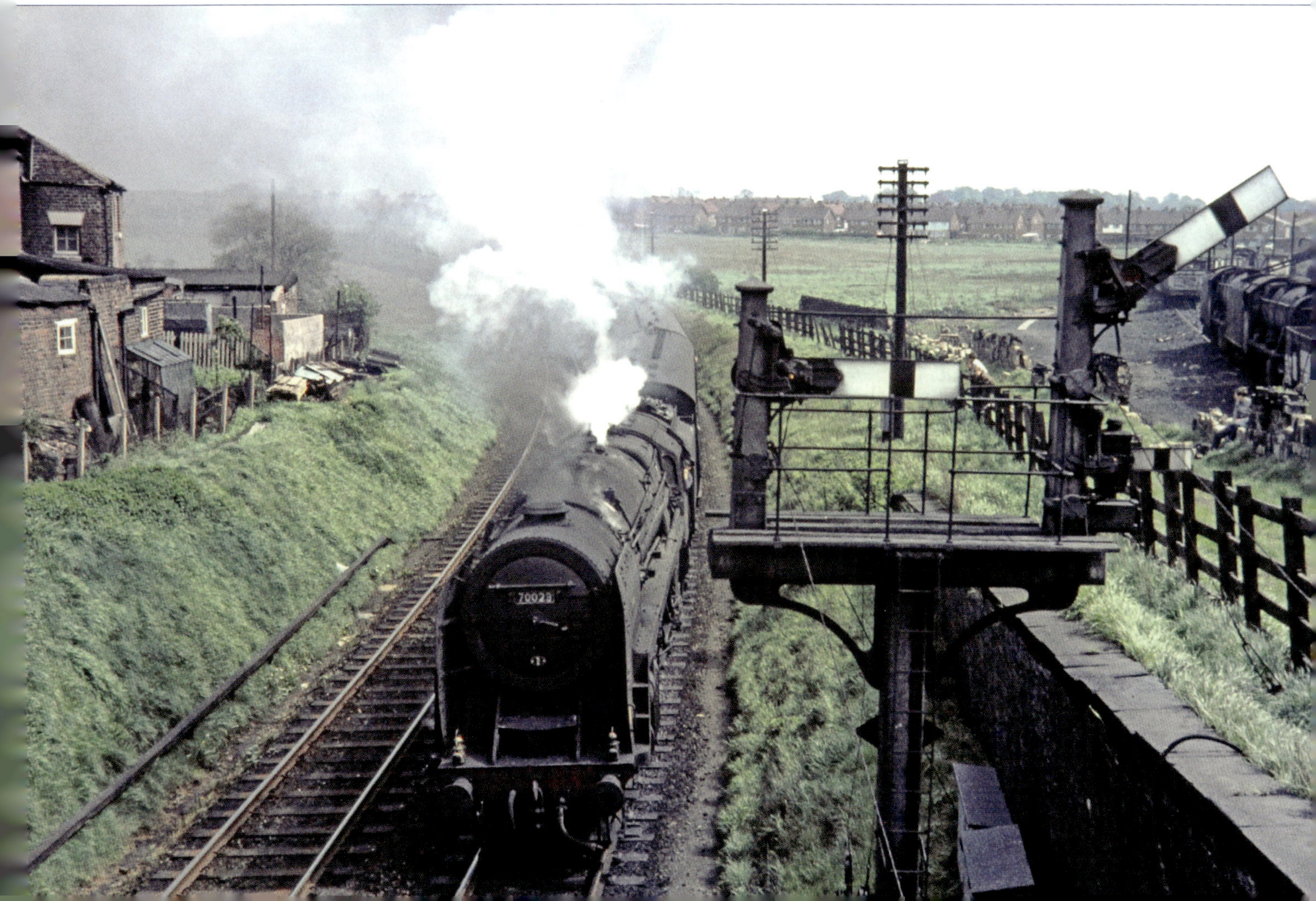

4 WEST CHESHIRE

Top left. The line built by the Birkenhead Jt Railway in 1863 from Hooton met its 1850 BLCJR line at Helsby Jct., thus giving Birkenhead and its port direct access to Warrington. On 18th June 1966 a very clean Standard class 9F 2-10-0 no. 92163 from Birkenhead shed emerged from the branch line onto the main (1850) line with a train of loaded tanks from Stanlow heading to Warrington and beyond. Moments earlier, Standard class Britannia 4-6-2 no. 70028 (formerly *Royal Star*) had swept through the station with the 11.25 Newcastle – Llandudno service. Helsby station building is unfortunately hidden by loco exhaust, but it has three platform surfaces, the central one a large V-shaped island platform which also still accommodates the LNWR type 4 signal box of about 1900. Only the station's name had seen paint recently. There was a small goods yard opposite the station's front door still with a line of wagons though the yard closed in 1964 so these were awaiting removal. The walkway linking the platforms was the standard LMS covered version, and there was a third stone waiting area on the far end of the branch line's platform 4. Also about to disappear was a small shed structure in LMS 'colours' at the near end of the same platform. Note the cross over (bottom left) from down main to branch line, and at the far end of platform 2 a turn out from the Up main gave access to two adjacent sidings and a loop which linked these sidings to the Birkenhead line. *HF-S644*

Bottom left. It's a summer Saturday morning (21st August 1982) and the removal since 1966 of the covered walkway helps to reveal the fine-looking structure in honey-coloured sandstone of Helsby Jct station. The 09.33 departure from Lime Street to Llandudno is just arriving behind class 40 no. 40124. One potential passenger is waiting on platform 1. Heading the other way is class 47 no. 47282 with a trainload of coal which may have been mined at Point of Ayr colliery (closed 1996) and be destined for Fiddlers Ferry power station. The class 101 DMU standing at platform 4 is the recently-arrived 09.31 from Rock Ferry, the Birkenhead terminus for all diesel services after the closure of Woodside and before the extension of electric trains to Hooton in 1985. At the station itself the rails to the goods depot have gone and a new road and housing development are about to get under way. The footbridge linking the three platforms is new, but the repeater home signals and their lofty signal posts have not yet been reduced. *IMG-0898*

Above. Helsby Hill rises abruptly 450ft (140m) above the town and gave this view over the station looking north on 26th August 1982. The photo shows class 47 no. 47100 heading towards Chester with a train of loaded Seacows. These are 40 ton, 4-axle ballast hoppers, equipped with controls at the ends of each hopper to dispense ballast centrally or to either side. A batch of them was built in 1981-2 and these look clean enough to be new. Beyond the railway is an area of marshland with some industrial development (there has been more since 1982) but no habitation. In the foreground the removal of station yard sidings has left an untidy mess of informal parking and preparation for development. The main station buildings, possibly faced with sandstone from the local quarry, show the same Dutch influence as at Ellesmere Port, and also have redbrick chimney stacks. On the island platform the signal box and building which houses waiting rooms both show well, but the waiting room on platform 4 has been removed. The large circular area at the far end of the island platform is unlikely ever to have had any railway function, and was probably a flower bed when there were staff available to tend it. *IMG-0905*

Above. Helsby Hill was also the vantage point for this wider view of the lines running north and west from the station, which occupies the bottom right corner of the image. The date is 10th December 1983. Leaving the station towards Chester is a type 4 class 45 ('Peak') loco probably on a Manchester Victoria – Bangor service. Alongside and just ahead of the 'Peak' are two lines of vehicles on the two sidings. Apart from some hoppers it is impossible to be sure what else these lines contain, but the further one has the front end of a class 40 loco just showing. Beyond these lines, on the loop which links the sidings to the Helsby-Hooton branch, another class 40 loco is at the far end of about 22 empty mineral wagons. Crossing the scene is the M56 motorway whose bridge passes over not just the Helsby-Hooton branch but also a line arriving from the left (marked by a string of vegetation) to join the branch at Helsby West Cheshire Jct signal box which is just beyond the M56 bridge. This is the 1869 CLC link from Mouldsworth on its Chester-Northwich route, which gave the CLC access to Birkenhead via running powers over the BJt. Beyond the M56 and to the right of the branch is a fertiliser works (far right) and Ince 'A' power station, while to the left under the tall chimneys is Stanlow oil refinery (all now closed). *IMG-0571*

Top right. On 28th September 1985 one of the quartet of class 40 survivors which were taken into departmental stock earlier that year, and said to have performed "almost faultlessly" in semi-retirement, was found lifeless at Helsby on one of the sidings adjacent to the Chester line. The offendor was no. 97406, previously no. 40135. After a while a pair of class 25s, nos. 173 and 201, arrived to haul the casualty away. The route was to drag the class 97 via the eastbound main line through the station, then to push it along the branch line as far as Helsby West Cheshire Junction where connection was made with the CLC link to Mouldsworth and then onwards no doubt to Crewe for attention. But before leaving the environs of Helsby the three locos now led by the two 25s passed the CLC's absurdly grand one-platform Helsby & Alvanley station (the photo unfortunately somewhat curtailed by a mishap in processing). The single track line with two intermediate stations (Manley the other) had opened for passengers in 1870 until 1875, then excursions in the 1930s, and finally workers trains from 1939 until closure of the stations in 1964. What mattered to the CLC (and BR) was freight on the Mouldsworth to Helsby West Cheshire Junction link from its opening for goods traffic in 1869 until its closure to all traffic in 1992. *IMG-1211*

Bottom right. Frodsham station has a sandstone cutting and tunnel at its NE end and an embankment and over-bridge at the other. It is therefore remarkable that sidings to both up and down lines could be squeezed in, but Frodsham did enough freight business that the main yard on the south side had a goods shed alongside platform 2 which is still there while its future use is debated. Summer Saturdays in the mid-1960s on the West Riding to North Wales route through Frodsham could generally be relied on to produce one of the few remaining Jubilee class locos. On the 20th August 1966 it was no. 45647 (formerly *Sturdee*) on a Leeds - Llandudno service which obliged, but shortly after came the unexpected appearance of Standard Class 9F 2-10-0 no. 92091 from Speke Jct shed (note '8C' on the buffer beam) hauling a two-car DMU. Light reflected from the sides of the cars suggests the pair had recently been in for attention and repainted, in which case they would hardly have needed to be delivered, but assuming they had failed the 9F would be taking them to Chester depot for repair. Unfortunately they passed non-stop. The two main losses to the station from the picture are awnings supported on iron columns to protect the up platform from the elements, and Frodsham signal box has gone from its place at the eastern end of the down platform. *HF-S926*

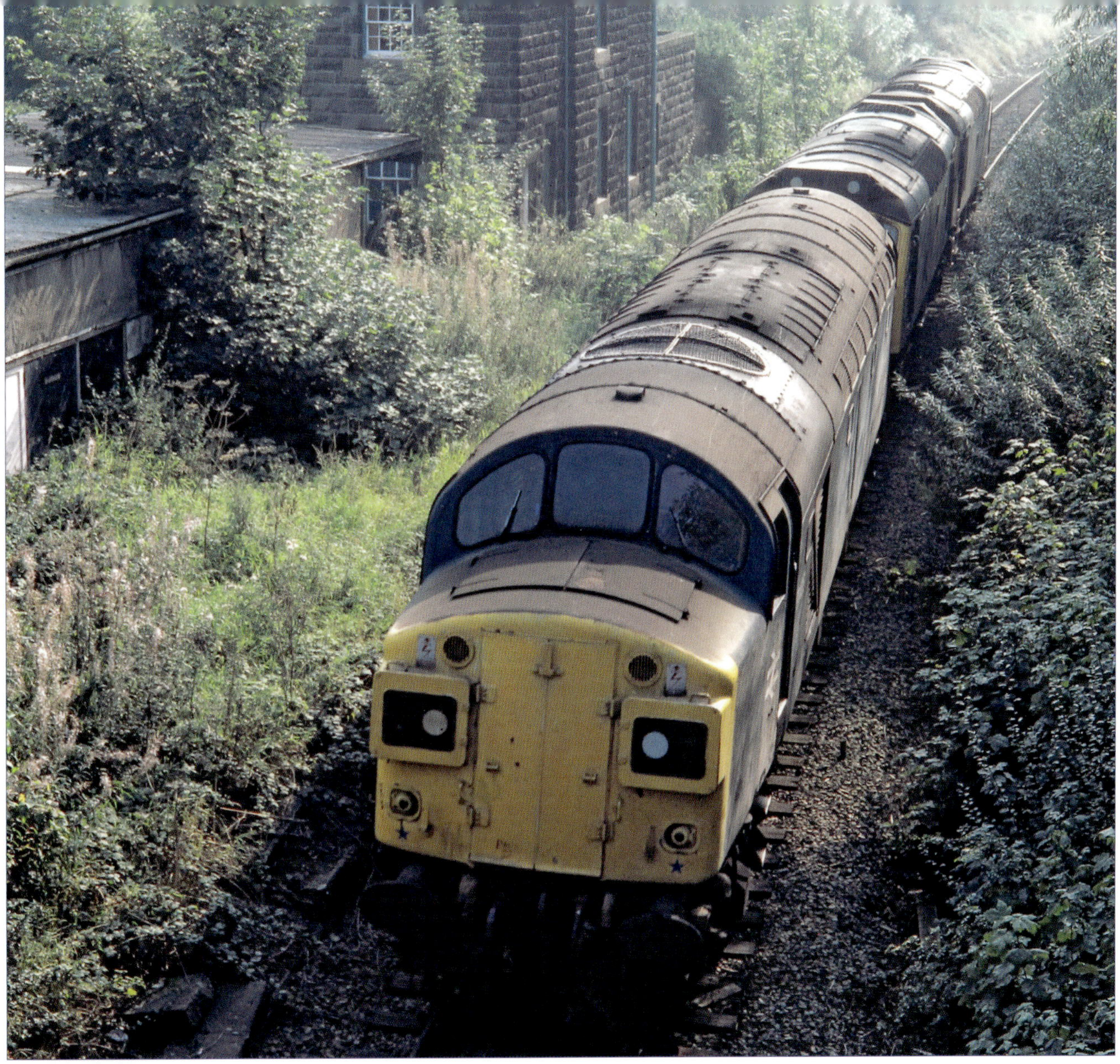

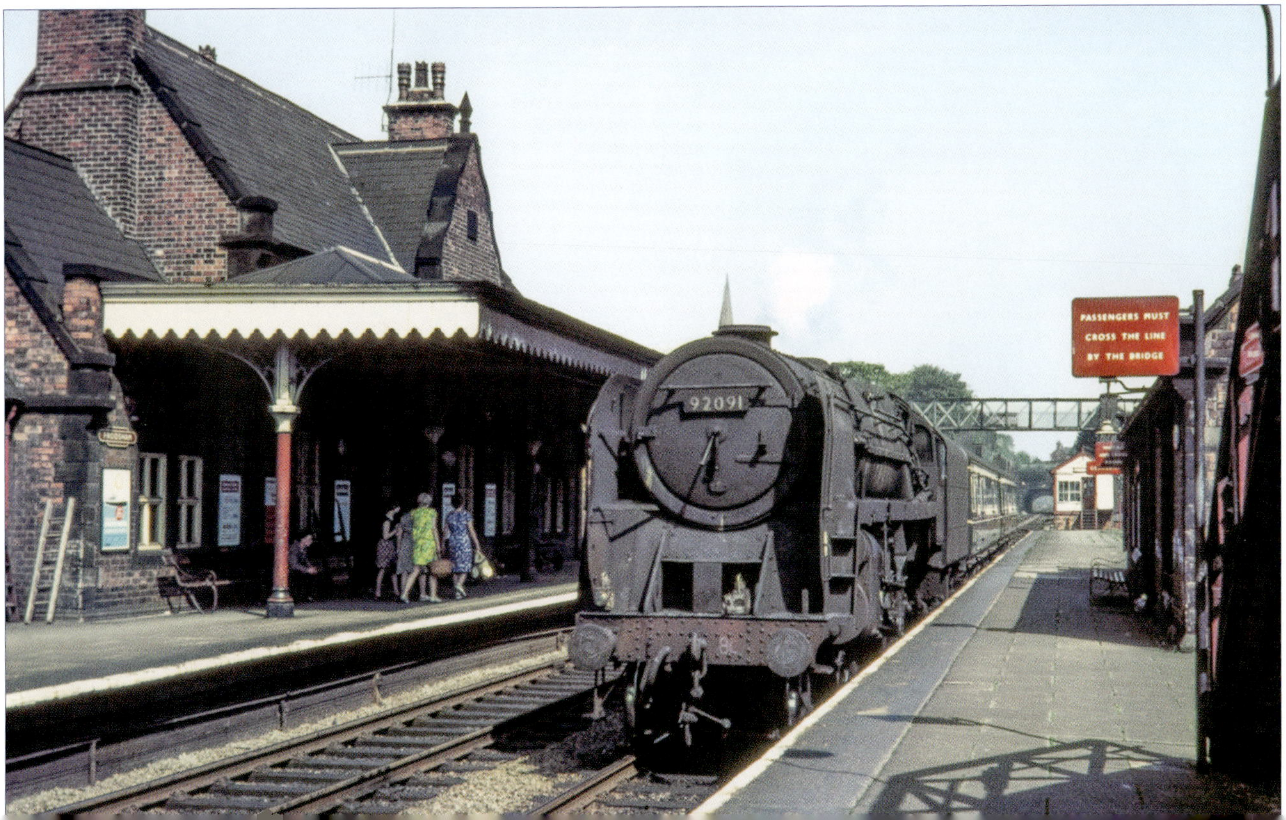
PASSENGERS MUST
CROSS THE LINE
BY THE BRIDGE
92091

74

Merseyside and West Cheshire Railways 1965-1990

Top left. The weather at the start of 1987 was very severe, and although the worst was over by the 17th January when this photo was made there is a nice contrast between the snowy landscape and the lighted carriages. With the loco's electrical heating for the train the passengers should be kept warm enough. At the head of the 15.16 service from Llandudno to York which has just arrived for its scheduled stop at Frodsham is class 45/1 1Co-Co1 diesel electric no. 45134. Built in 1961 as D126, the loco was one of the last 45s to be withdrawn, in September 1987. The first vehicle bears the new livery and name of Regional Railways with its windows revealing a noticeably different light. The station awnings over platform 1 have disappeared, as has the Frodsham signal box which stood in the immediate foreground just beyond the end of platform 2. The curved retaining wall on the right marks where the up sidings were located. That space was already in use for car parking and the same soon happened on the left side. The station building stands out well against the pinkish evening sky, and the line of light in the far right distance marks the route of the M56. *IMG-1593*

Bottom left. A mile or so east from Frodsham station the Chester-Warrington line crosses the valley of the River Weaver which is spanned by a quarter-mile viaduct. The main feature of the viaduct is the bridge over the river itself which has two cast iron main spans, followed by 21 stone arches. (At the far end of the valley there is a similar single span iron bridge over a canalised stretch of the river as the Weaver Navigation.) Parallel to the railway is the old Sutton Causeway which became the A56, whose bridge over the river gave attractive views of Sutton Mill and Frodsham Wharf. It still does, but in 1966 before the M56 was built the A56 on summer Saturdays carried a constant stream of slow-moving exhaust-emitting traffic. Midday on 20th August was hot and airless, so after a busy morning I dived into the adjacent Bridge Inn for a quick refreshment and emerged just in time to grab one of Stanier's class 8F 2-8-0s crossing the river towards Stanlow with a train of Shell oil empties. Only afterwards did I realise my good fortune in the mirror-like surface of the water, the motionless swans, and the perfect trail of loco exhaust with only the passage of the train itself to disturb it. *HF-S924*

Below. On 19th August 1985, a class 47 loco was heading towards Warrington with loaded HAA coal hoppers probably for Fiddlers Ferry power station. From oil hauled by coal in 1966 we had graduated to coal hauled by oil in 1985. Below the bridge the 1966 image of tranquillity at Sutton Mill and Frodsham Wharf was transformed. It appears that in the early 1980s a trade in grain developed involving its transfer at Seaforth Grain Terminal to small ships, and its further transfer at Frodsham Wharf for onward movement by road. Four of these vessels are lined up with the yellow grab crane shifting the grain to the grey HGV. From left to right, the smallest of the boats is not showing any ID, but next to it is the *Panary* of Runcorn, built in Bristol at the Charles Hill yard in 1937. *Panary* is listed in the National Historic Ships Register and was a frequent visitor to Frodsham at this time. The largest of the group is the *Spurn Light* which may have been a sister ship to the *Humber Trader* (built Hull 1957) which had certainly been active on the Mersey and Weaver. Finally the *Parcastle* was a Duker barge built by Pimblotts of Northwich in 1952 which (I read) in retirement in 1993 was renamed *JD McFaul* and was sailed across the Irish Sea to Shannon. *IMG 1990*

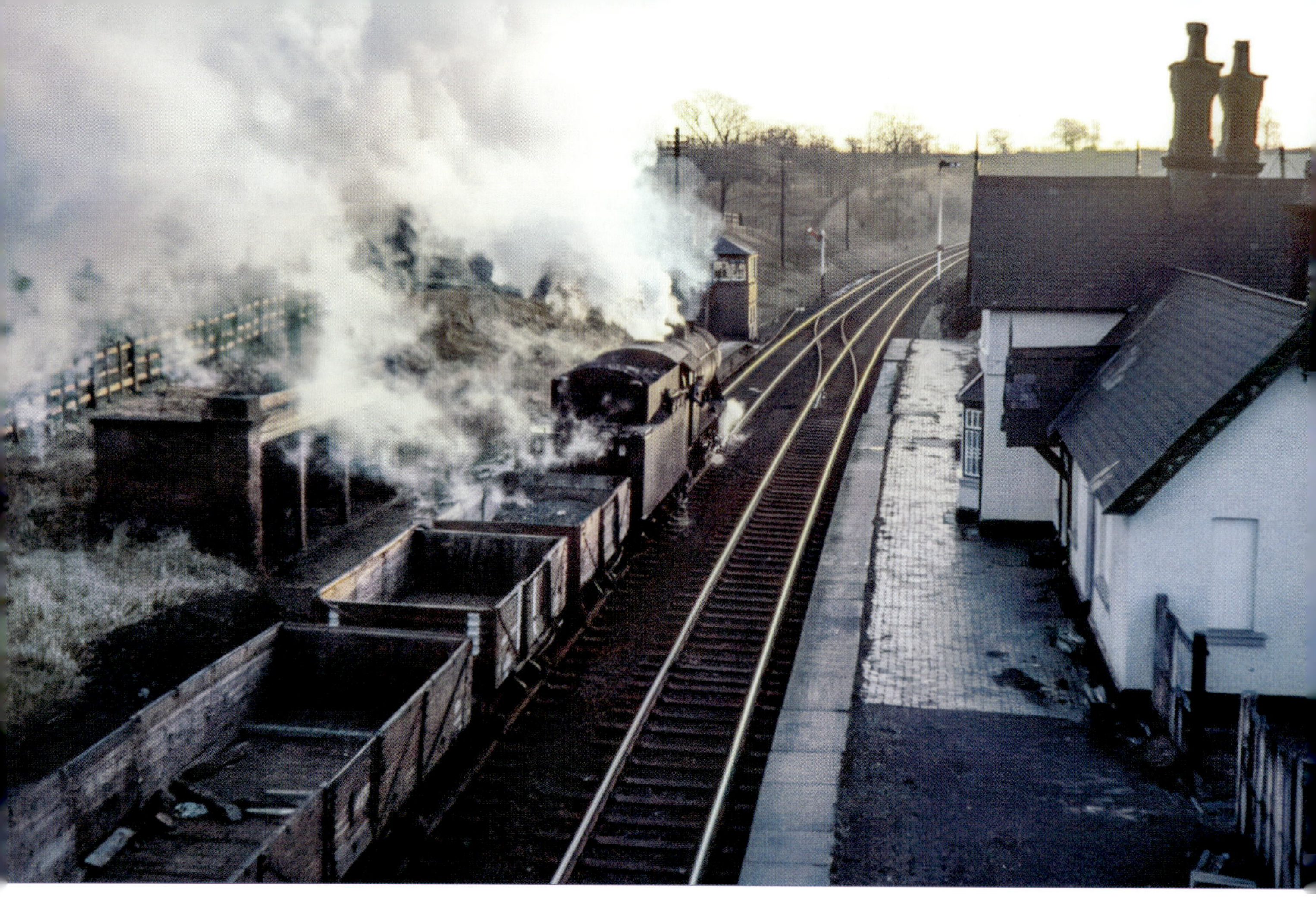

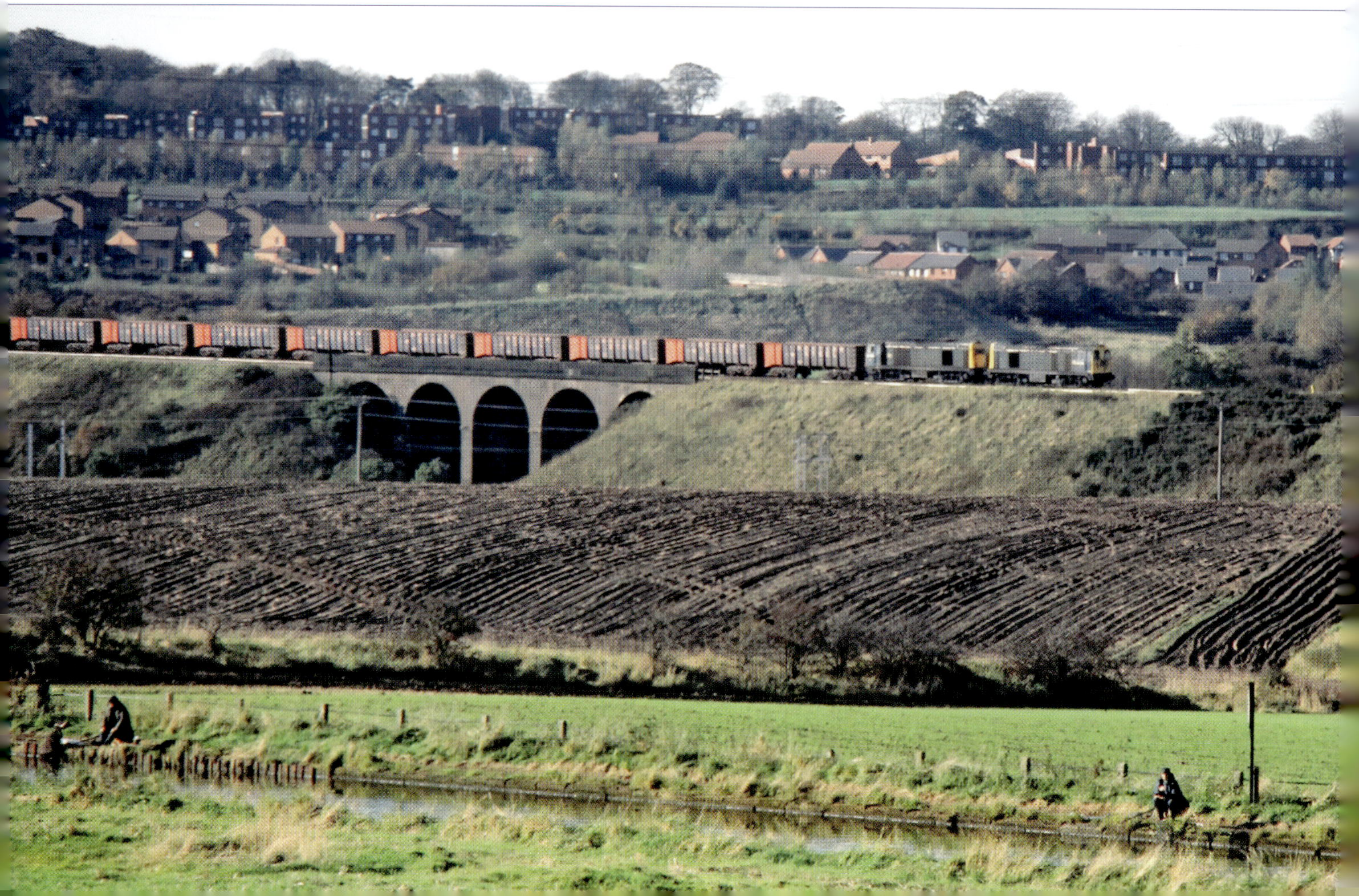

Top left. About 2 miles north-east from Frodsham the 1850 Chester-Warrington line enters Sutton tunnel, while above ground the 1869 Runcorn-Weaver Jct line passes over it but makes no connection other than the Halton Curve to Frodsham Jct. When the BLCJR built the tunnel in 1850 a station was located at each end. Halton station was only 1½ miles from Frodsham and had some limited capacity for handling goods traffic, but it was closed in 1952. Norton station closely resembles Halton (both are still extant) but it too closed in 1952. The photo shows WD class 2-8-0 no. 90113 passing between Norton's platforms, catching the last of the day's sunshine on 2nd February 1966 as it attacks a stretch of 1 in 200 on the north-eastern approach to the tunnel hauling a mixed collection of empties. 'Norton' was in reality an area rather than a place, but it has been comprehensively absorbed into Runcorn New Town and a new station and signal box for Runcorn East were opened in 1983. Their location is just beyond the trailing crossover on the westbound track in the photo. Ironically, Halton had previously had the closer connection to Runcorn, being named Runcorn Road or just plain Runcorn until the name was needed by the LNWR in 1869 for its new facility on the south shore of the Mersey. Loco no. 90113 was a Wakefield engine but Warrington Dallam shed was making use of it again three months later. *HF-S98*

Bottom left. On 29th October 1988 a pair of class 20 locos hauling a rake of bogie (four-axle) wagons was seen about a mile beyond Runcorn East heading towards Warrington. They were crossing a five-arch bridge over the WCML and Keckwick Brook below, with the tops of the masts just visible on the left. The wagons look new or recently painted with 'British Steel' prominently displayed. These are PTA tippler wagons, originating in the 1970s and would normally be described as 'rugged' or 'battered' rather than 'pristine' from handling anything (in the steel context) from iron ore to dolofines. Shotton Steel Works on Deeside (John Summers Ltd until 1967) was just the other side of Chester and the nearest British Steel works in 1988, so it would be easy to assume this was a routine working from one BS works to another. But the ten wagons (perhaps more) appeared to be empty and not causing problems for the two class 20s, so it seems as likely that they had been in for maintenance and a coat of paint somewhere and were being delivered to where they were needed. Meanwhile, in the foreground of the photo is a stretch of the Bridgewater Canal and two anglers ignoring the railway. *IMG-1111*

Above. In 1861 the LNWR received parliamentary approval to build a bridge over the Mersey at Runcorn Gap, which when completed carried its first revenue-earning traffic in 1869. The bridge, and the 8½ miles of approach to it from north and south, directly connected Liverpool to Crewe and hence to the LNWR's Euston terminus in London. The bridge was built in three sections of wrought iron, each 93m long, supported on two abutments in the river. The structure consists of an iron floor supporting the double track line with lattice trusses supported by box girders to the abutments. The height of the bridge, 75ft (23 metres) above high-water level, was required for the passage of shipping on the river and, later, on the Manchester Ship Canal. The turrets and other structures are built in the style of Gothic castellation. The bridge is called simply the Runcorn Bridge, but for some it is the Britannia Bridge after the image on the painted shields or even more fancifully the Aethelfreda Bridge for a local Saxon princess. This view of the bridge seen from Runcorn, framed above by underparts of the adjacent Silver Jubilee road bridge (1961), and below by the Ship Canal, was made in November 1985. In the distance beyond the central section of the bridge an HST is just visible. *IMG 1218*

Merseyside and West Cheshire Railways 1965-1990

Top left. On the 15th January 1966 this Stanier 8F 2-8-0 was working hard on the northern approach to the Runcorn bridge. The climb begins at Ditton Jct at a gradient of 1 in 114 for close to two miles in which the train negotiates a curve of about 110°, from heading east to roughly south by west. The 8F is still on the initial embankment but is approaching the first section of viaduct – 49 arches – to be followed by a short resumption of embankment and the second viaduct section of a further 16 arches. The goods train in the foreground has arrived from the Widnes Deviation Line and is diving through the embankment to continue westward. The crumbling high level abutment on the left is a remnant of the Ditton Marsh Branch of the joint GCR-Mid. line built in 1878 to connect with the CLC main line at Hough Green Jct. This section of the Branch was severed in 1960 (the rest closed in 1964) when the 1961 road bridge was built. When I revisited the scene in October 1982 the six rail sidings on the left were reduced to two, and of the two signal posts in 1966 everything to the right of the left-hand post (including the original SHCRC route and the three adjacent sidings) had gone. *HF-2-3-32A*

Bottom left. On my first visit to Runcorn on 15th January 1966 I walked north from the station on the road bridge admiring the railway bridge, regretting the way its tight lattice work obstructed the view of trains as they crossed, and seeking a good location to photograph the bridge approaches. The day was damp and misty so not great for photos, but I was curious to see how close I could get to the line. Soon enough I found a long series of steps up from the road to the line side with no physical obstruction or warning notice to indicate (of which I had no knowledge) that as from the previous year the footway over the railway bridge was permanently closed to all but BR maintenance staff. In my innocence I settled to enjoy the spectacle of Standard class 9F 2-10-0 no. 92048 with a long freight beginning its climb in the far distance, soon to be appreciated at very close quarters. Over to the left was the West Bank Dock Estate with its railway system and the Docks themselves all about to be abandoned. Fisons was in fact the last to lose its rail connection in 1981. *HF-2-3-30A*

Above. A few minutes after no. 92048 disappeared onto the bridge it was followed by one of the first batch of 25kV electric locos of the AL1-AL5 series (TOPS classes 81-85). As it passed it left me with a vivid memory of some stuff dropping from the loco's pantograph landing with a splat more or less at my feet. I guess it was simply an accumulation of soot dislodged from the overhead cable but I was relieved it had missed me. The last loco to cross while I was there was a class 8F 2-8-0 no. 48711 running light engine which emerged northbound from the bridge in a cloud of steam. Curiously the loco seemed to be paired with a Midland style tender, probably from a 'Crab' or Stanier 2-6-0. Deciding to quit while I was ahead I succeeded in crossing the bridge to the Runcorn end unchallenged, without having to endure the noise of a passing train on the bridge. Nor did I have to pay the toll that was demanded of previous generations though it would have been well worth it. *HF-2-6-1*

Above. On the same day and nearly from the same spot as HF-2-3-32A (p.79), this is a glimpse of how things used to be on the connection of the railway to the chemical and metal-working industries which dominated the town of Widnes for over a century, and have now almost all disappeared. McKechnie Brothers Sulphate of Copper Works was established in Widnes in 1891. The company built its HQ offices in Ditton Road Widnes in 1920 (to be demolished in 2021). On BR rails an 03 class small diesel-mechanical shunter no. D2372 (built 1961, withdrawn from service in 1970 before TOPS 03 numbering was introduced) from Speke Junction shed was at work. On the far side of the gate is an active steam loco which appears to be Andrew Barclay 0-4-0 saddle-tank named *Efficient* (works no. 1598 of 1918). This loco was the last of four steam or Sentinel engines employed throughout its working life at McKechnie's. It was acquired privately in 1969 and is now housed at Ribble Steam Railway and Museum in Preston. By 1986 all the BR infrastructure in the picture connecting the works to the railway system had gone. *HF-2-14-11A*

Top right. In the centre of Widnes and just north of Widnes Dock Jct was a long-established footbridge which on 3rd December 1965 provided a convenient location to view railway activity. This, the first of two photos from that day, is a view looking northward of the St Helens and Runcorn Gap Railway (SHRGR) which reached Widnes in 1832. Occupying the track while shunting a 20-ton 'Presflo' cement wagon is an Ivatt 2-6-0 loco no. 46410 from Speke Jct shed (Widnes' own shed 8D closed in 1964). The track disappearing off to the west was a spur to the 'low-level' line built in 1853 by the St Helens Canal & Railway Co. (SHCRC) to link Warrington with Garston. (The SHCRC was formed in 1845 when the SHRGR amalgamated with the St Helens Canal / aka Sankey Brook Navigation whose New Cut reached Runcorn Gap in 1833.) The two lines, roughly N-S and E-W, crossed on the level at Widnes Dock Jct. The resulting congestion was reduced by the LNWR's Widnes Deviation Line of 1869 which is just visible as the bridge over the 1832 line in front of the gasometer, but the crossing remained intact until probably 1969. *HF-S18*

Bottom right. From the footbridge looking south is the SHRGR's Widnes dock of 1833, about a half-mile distant, which gave access to the Mersey for coal from St Helens. The loco on the dock line is 8F 2-8-0 no. 48033 from Birkenhead's 8H shed. Three lines appear from the left or bottom of the photo: a spur which joins under the loco's brake van links the 1853 line from Warrington to the 1833 dock line. A second line crosses both tracks of the 1833 line in front of no. 48033 but connects to an industrial line, probably a minor inconvenience. But squeezed up in the bottom right hand corner of the photo is just a few feet of the 1853 line crossing the 1833 line, the main cause of congestion. The four large objects are a mystery. The pale silver-grey strip is the St Helens Canal which the railway crossed by a swing bridge, the canal reaching the Mersey to the west of the dock. Beyond the bridge over the canal another Ivatt 2-6-0 loco is at work, and beyond this loco four landmarks (L to R: St Mary's church tower, the 1959 road bridge, the 1860 Tower Building / Catalyst museum, and two turrets of the Runcorn rail bridge) are faintly visible in the murk of a December afternoon. Apart from the Deviation line the railway in Widnes has long since vanished, leaving Spike Island and the canal as recreational areas. *HF-S19*

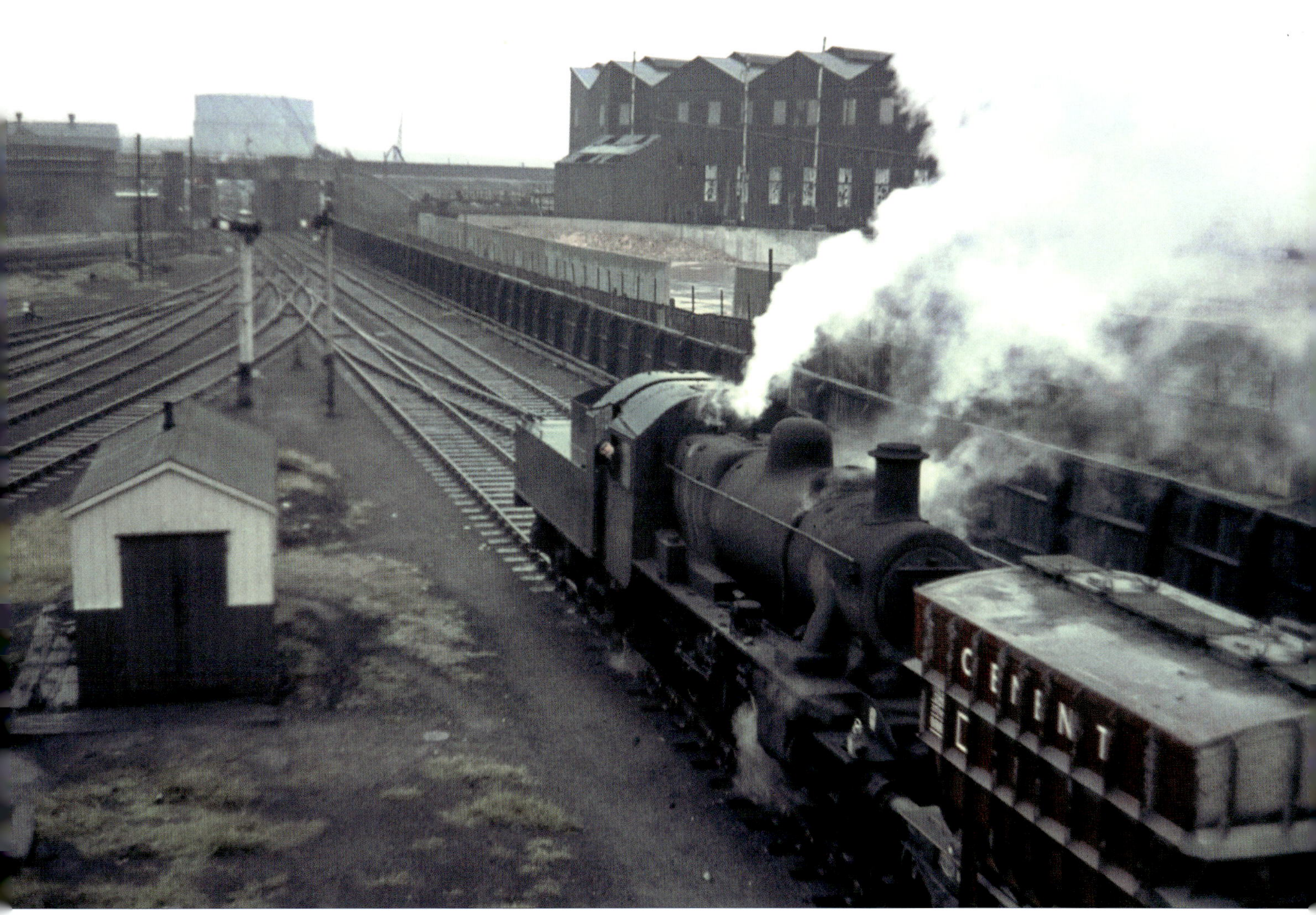
CEMENT

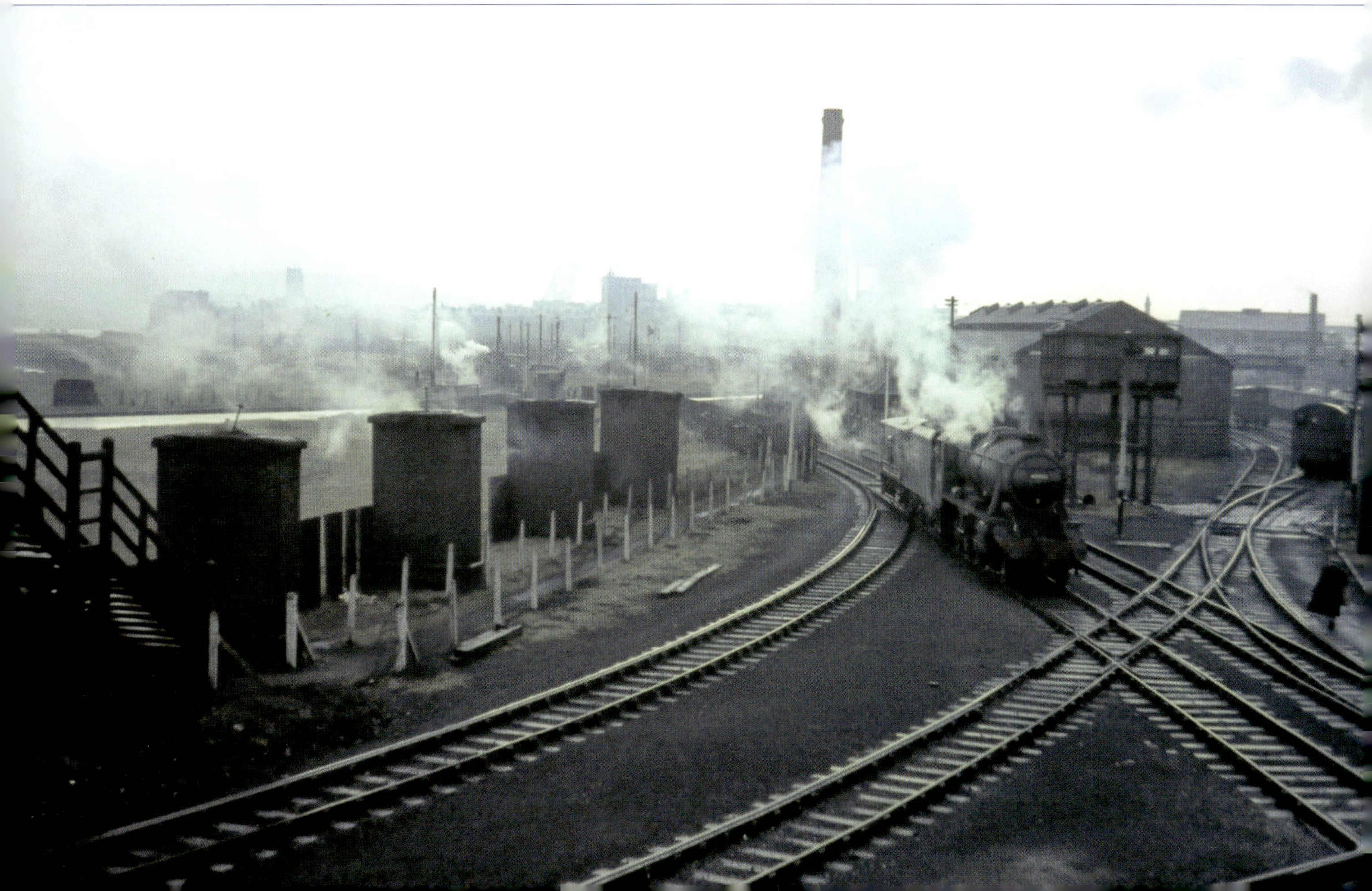

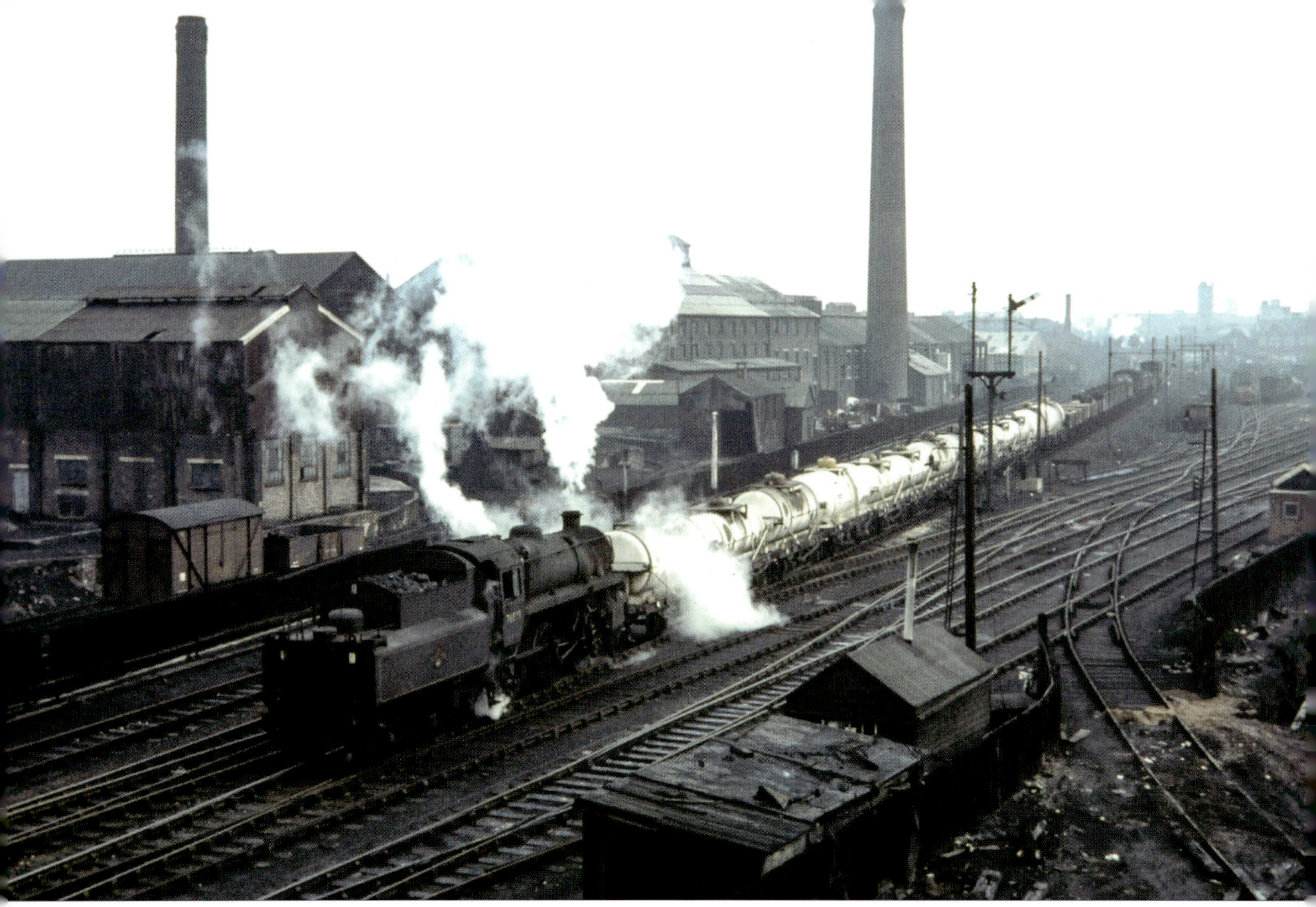

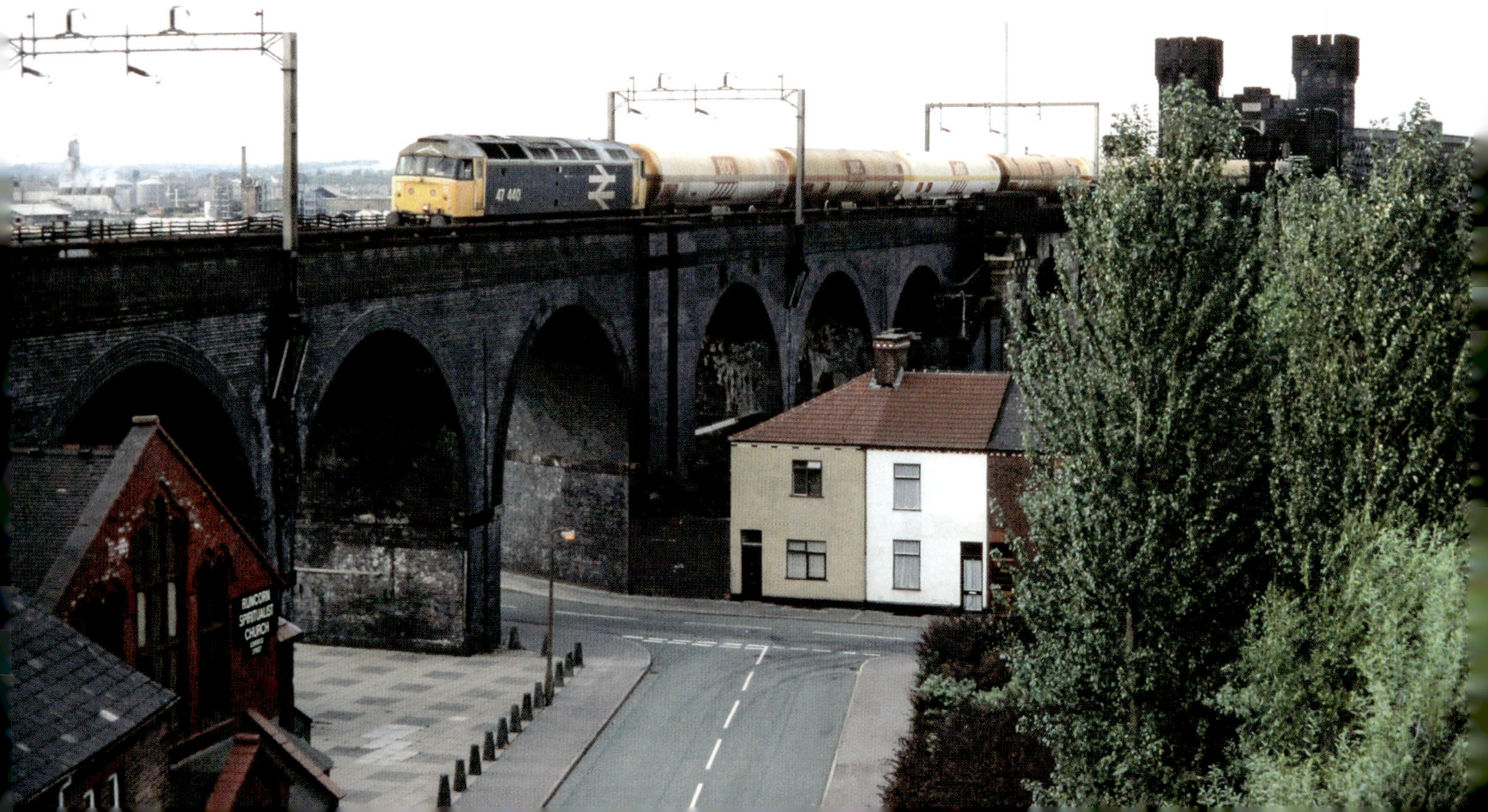
47 440
RUNCORN SPIRITUALIST CHURCH

Top left. The original 'low-level' route between Warrington and Garston was built by the SHCRC in 1853 to develop an expanding trade to Garston docks, but movement through Widnes was impeded at two 'pinch-points', the crossing on the level of the two main lines, and a road level-crossing at Waterloo Road in the town. When the LNWR secured control of the SHCRC it built the Widnes Deviation Line, a 1½ mile 'bypass' which opened in 1869. The Deviation was built to the north of the original line, from Carterhouse Jct in the east, to Widnes West Deviation Jct from where this photo was taken on 5th November 1966. Standard class 4 2-6-0 no. 76079 was coasting tender-first down the embankment from the bridge over the St Helens line on the 1 in 97 slope with a long train mostly of tanks. In the bottom right corner is an industrial line which crosses three sidings to reach the original SHCRC route. Out of the picture to the right are the Hutchinson Street sidings and goods depot, and faintly in the distance are further sidings between the Deviation and the original line. *HF-S1356*

Bottom left. When the Jubilee road bridge alongside the Runcorn rail bridge was built in 1961, it included a spur leading for a short distance from the northbound lane of the road on its approach to the bridge. Protected from traffic on the road, this spur gave a safe and commanding view of the southern end of the rail bridge. On 27th August 1988 class 47 no. 47440 was exiting the bridge with a train of British Oxygen Company tanks, all looking as grubby as the loco. The BOC operated its Ditton works on the northern side of the Mersey from 1970 until about 1998 when the site was taken over by the Widnes Intermodal Rail Depot. The Runcorn Spiritualist Church in Ashridge Street is well-known for its location, its entrance projecting from under one of the railway arches. The rear of the building in Brindley Street is more conventionally flush with the stone-work of the arch though unattached to it. The building and its location date from 1888 but was acquired by the Spiritualist Church from a Methodist Church in 1906. James Brindley as engineer of the Bridgewater Canal is also commemorated in Runcorn by the Brindley Theatre in the town. *IMG-1107*

Above. On 4th October 1985 looking south towards Runcorn from the spur on the road bridge, a self -propelled work train appeared on the down line. I later learned that this was a Speno Rail Grinding Train, manufactured in Switzerland, of which two sets of the five-unit trains (a motorised driving unit at each end and dormitory etc facilities in the middle three) had been on hire to BR probably since the early 1980s. It was Speno's practice to hire out the operators (mostly Italian) as well as the trains, though employing a BR driver. One of the sets worked exclusively on BR's LM region, and was reported to be parked at Birkenhead loco shed around the same time. Long since replaced, both units were found in March 1990 near Sheffield waiting to be broken up so presumably Speno did not want them back. In the foreground, the Devonshire pub has gone out of business like so many others but the building has been carefully 'rehabbed' and has a new occupant. The pale-coloured Runcorn and Widnes Co-operative Society Ltd (1928) building across the road is now the Concrete mango Creative Temple. *IMG 1214*

83

Merseyside and West Cheshire Railways 1965-1990

Merseyside and West Cheshire Railways 1965-1990

Top left. Between Runcorn station and the bridge but underneath the railway the Bridgewater Canal used to proceed to a sequence of locks that carried the canal down to the Mersey and later the Manchester Ship Canal. Building the Silver Jubilee road bridge in 1961 – specifically its slip roads – brought the canal to an abrupt end at Waterloo Bridge on Station Road. The 17th June 1984 was a warm mid-summer Sunday so lots of people were out taking the air, fishing or "simply messing about in boats" in the canal basin. But on the railway Sundays meant an opportunity for maintenance work, and type 4 class 45 1Co-Co1 no. 45048 *The Royal Marines* (still named and badged) had arrived at Runcorn taking possession of the down track. As well as a mess vehicle and tool van, the engineers had a special unit in the customary yellow livery for track maintenance, with arms on the unit for lifting that I could see were deployed to the 'safe' outside of the track. This evidently freed up the up line as a DMU was allowed to pass towards Runcorn station. Down by the canal the Waterloo Hotel must have done good business that day, but although it closed in 2012 the building has survived and is now a Buddhist Temple. *IMG 1382*

Bottom left. Passing through Runcorn station on 1st July 1982 is a diesel-electric type 2 class 25 Bo-Bo no. 25242 with a down train of empty Cartic-4 car transporter wagons. The loco was one of over 300 of the class built between 1961 and 1967. The Cartic-4 was developed in the 1960s, with Ford testing the prototype which resolved the problem of moving large numbers of cars between factories or to distributors. The double-decked wagons were built between 1966 and 1973 and were quickly adopted. The main problem which emerged in the 1980s was the vulnerability of new cars to stone-throwing vandals and thieves who targeted fitted radios, hence different types of screen were added to the basic structure. The train pictured here, as yet with no screen, was certainly returning to Halewood where Ford built a factory on new ground and began car production in 1963, being especially associated with the Escort model from 1967 to the end of the century. Runcorn station was built in 1869 together with the rail bridge over the Mersey. It was endowed with a small goods yard which has now provided the space for a four-level car park. The station itself has replaced the open footbridge to platform 2 (left in photo) with an enclosed bridge incorporating lifts for passengers and luggage at the far end of the station. *IMG 1320*

Below. Runcorn signal box at the south end of the station, seen here on 24th December 1983, was listed thirty years later as a Grade 2 building. Its merit is to be representative of signal boxes designed by the LMS at the beginning of WW2 in line with specifications laid down by ARP (Air Raid Precaution). It was one of the first to open, in January 1940, and has retained its original distinctive metal framed windows. In spite of their differences in size, location and date of opening, the Runcorn box is very reminiscent of Lime Street's own, each with 14-inch brick walls and 12-inch reinforced concrete roofs. Pictured here is EE type 4 class 40 1Co-Co1 no. 40079 standing short of the Runcorn box and adjacent to platform 2 but separated by a fence, waiting to be cleared to cross the down main line to reach the up main. It has arrived from the Folly Lane sidings of ICI's Castner-Kellner works with its train of TTA tanks on hire from PROCOR and BRT carrying caustic soda. These trains ran until around 1990 but declining business from the works plus BR's decision ahead of privatisation to wind down the Speedlink Network which had catered for wagon-load freight spelt the end for this traffic. *IMG 0585*

Merseyside and West Cheshire Railways 1965-1990

Below. With the Mersey crossing completed in 1869, the LNWR wasted no time in connecting Runcorn to the former GJR, and to the Runcorn docks where the Weaver Navigation reaches the Mersey and the MSC. The Folly Lane branch connects directly to the down main line at Runcorn station, having described a wide arc from Weston Point (the centre of the dock complex) to avoid Runcorn Hill. With no turning facility at either end of the branch, the loco seen here – class 8F 2-8-0 no. 48503 on 19th June 1967 – had earlier run up light engine from Speke Jct over the bridge to Runcorn Jct beyond the station to gain the down line, and thence tender-first down to the docks to pick up its train of 'Presflo' cement wagons. The loco has its shed code 9F painted on the smoke box door, unnecessarily unless the cast iron shed plate had been removed by a souvenir hunter, as Heaton Mersey (Stockport) shed had been the loco's domicile at least since 1951. The photo, taken from the old footbridge at the station, shows a few feet of track diverging from under the loco as it passes. It is hard to imagine what the purpose of that track would have been (or how far it went), unless it might hold a loco to move onto the rear of a freight arrival off the branch held on the down main line to then proceed northward. *HF-S1681*

Top right. On a bright and breezy 6th March 1985 I was alongside the sidings of the ICI Castner Kellner works, with the Hale shoreline on the other side of the river, and encountered this busy scene featuring class 08 diesel electric 0-6-0 shunter no. 08846 sorting the TTA tankers. Also hidden in the picture is a class 40 loco, and on the far side of the yard are what appear to be wooden bogie wagons (Procor TOPS code PNA?) one of which has the ICI emblem. A number of British loco builders began producing diesel shunters in the 1930s but the LMS led the way building in quantity during and after WW2. The other railways produced some of their own which I first encountered as ex-LNER BR-numbered 15000 series on a school visit to March (31B) shed and Whitemoor Marshalling Yard in about 1953. BR began building their own in 1952, many of which lasted long enough to be included in the TOPS numbering scheme, and over a thousand 08s (or similar 09s) had been on the books when building ended in 1962. No. 08846 was built at Horwich in 1961 as D4014 and was withdrawn from BR service in 1989 as BR's need for shunters declined. But no. 08846 had plenty of life left, and it was sold to Railway Support Services at Neville Hill (Leeds) from whom it was hired by Network Rail to work at its Rail Recycling Centre at Whitemoor (March) in 2022. *IMG 0696*

Bottom right. Weaver Jct is best known as Britain's and the world's first flying junction. Once the Mersey bridge was built in 1869 the LNWR immediately connected the new Runcorn station to the GJR line – by then absorbed into the LNWR – by a flat junction at Birdswood. But to minimise disruption to traffic, in 1881 the up (southbound) line was separated from the down and carried over the GJR (now WCML) which it then joined at speed a mile south of Birdswood. Here on 13th May 1989 EMU class 304 unit 40 of 1960-1 vintage is passing with the 08.44 Lime Street – Crewe. The slam-door trainsets lasted well but access via the seven doors per vehicle-side could be awkward past two sets of knees when the service was busy. Notice to the left, overhead equipment of the WCML at a lower level. Minutes later the 09.14 Lime Street – Cardiff went by with class 47 no. 47621 *Royal County of Berkshire* looking very regal in its recently-applied InterCity livery, but by then the black clouds had blocked out the sun. 'Weaver Jct' became notorious in the late 1960s as a shorthand reference to delays in obtaining approval for electrification beyond the Junction to Glasgow, eventually reached in 1974. *IMG-1664*

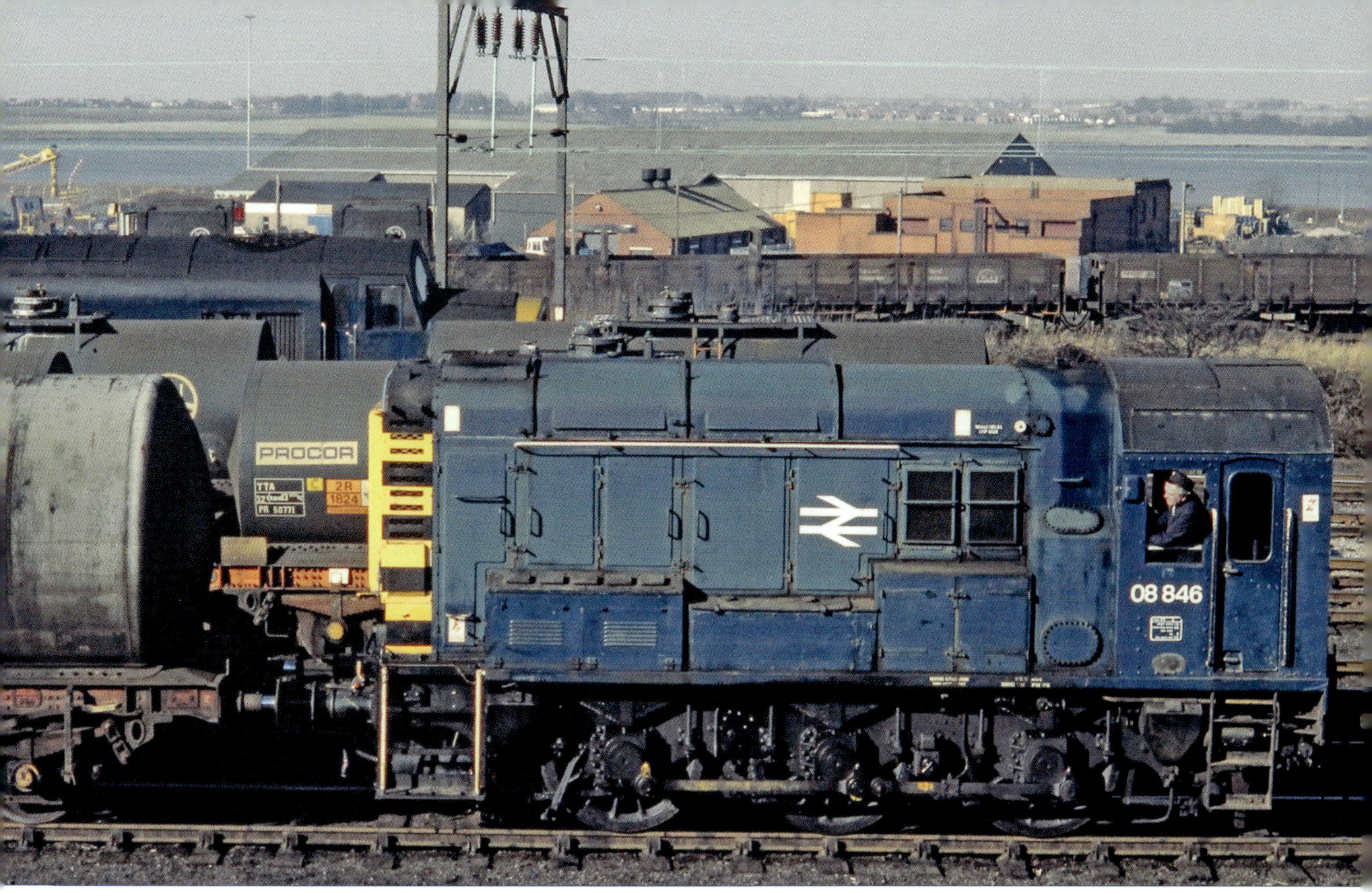
PROCOR
08 846

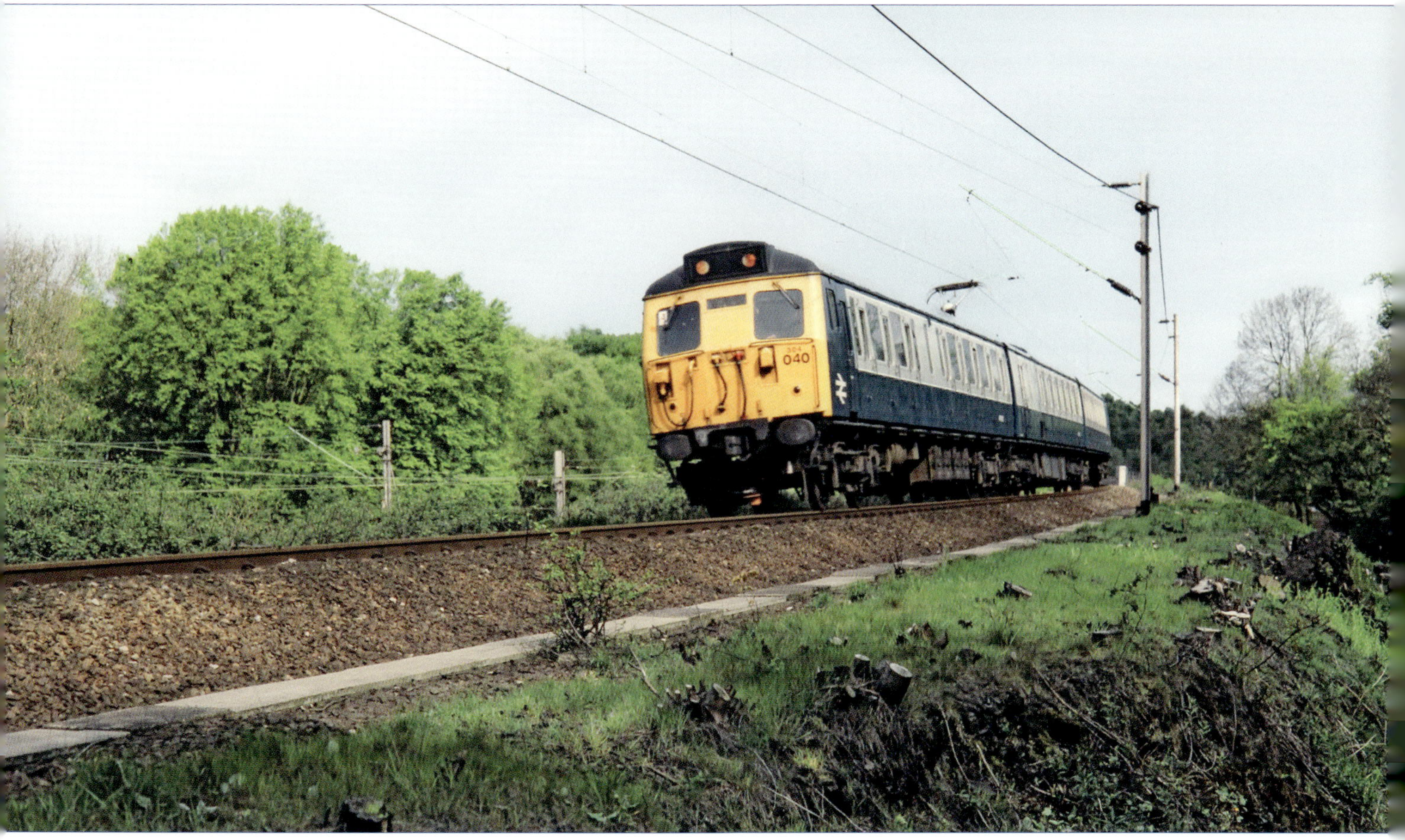
040

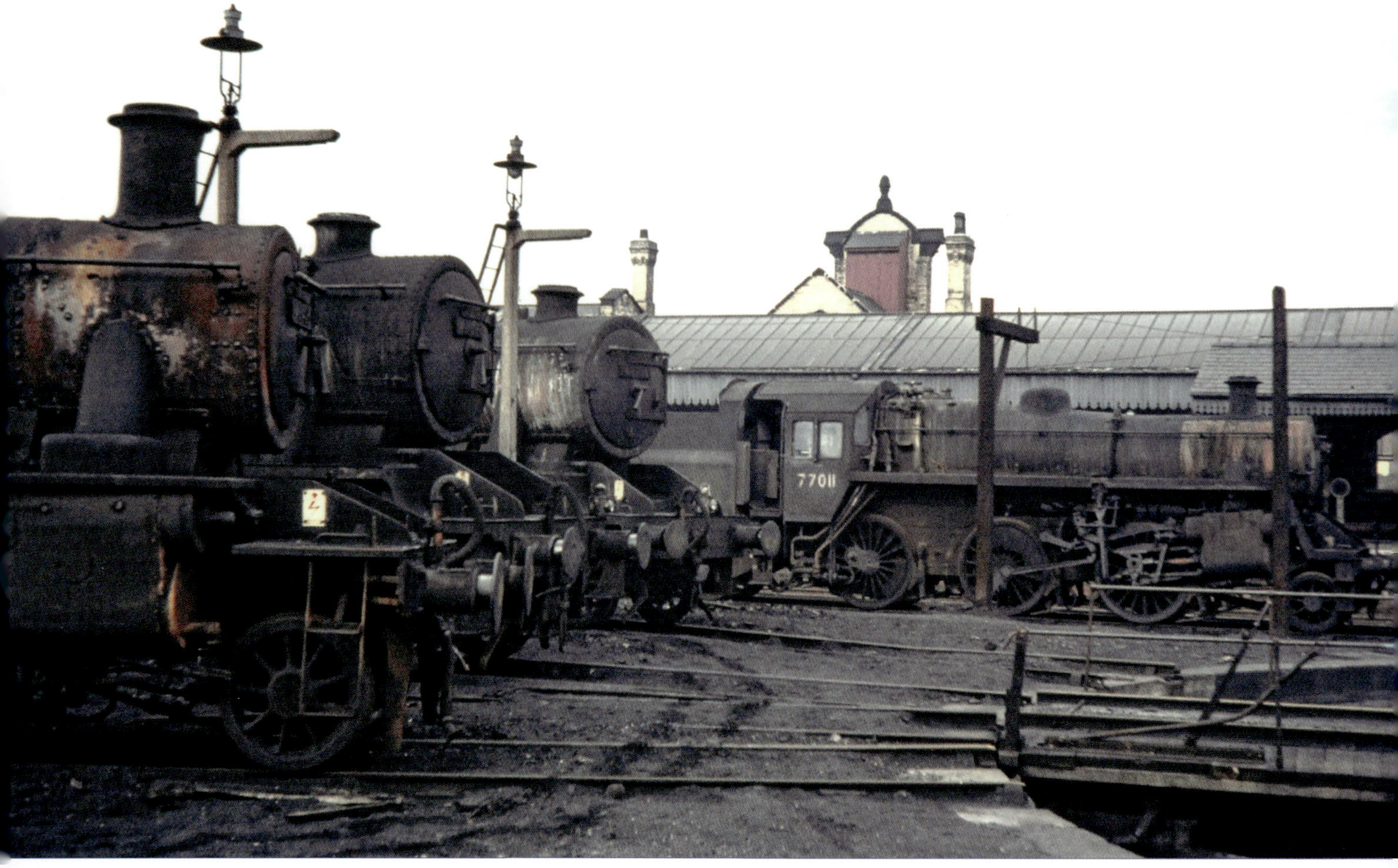
77011

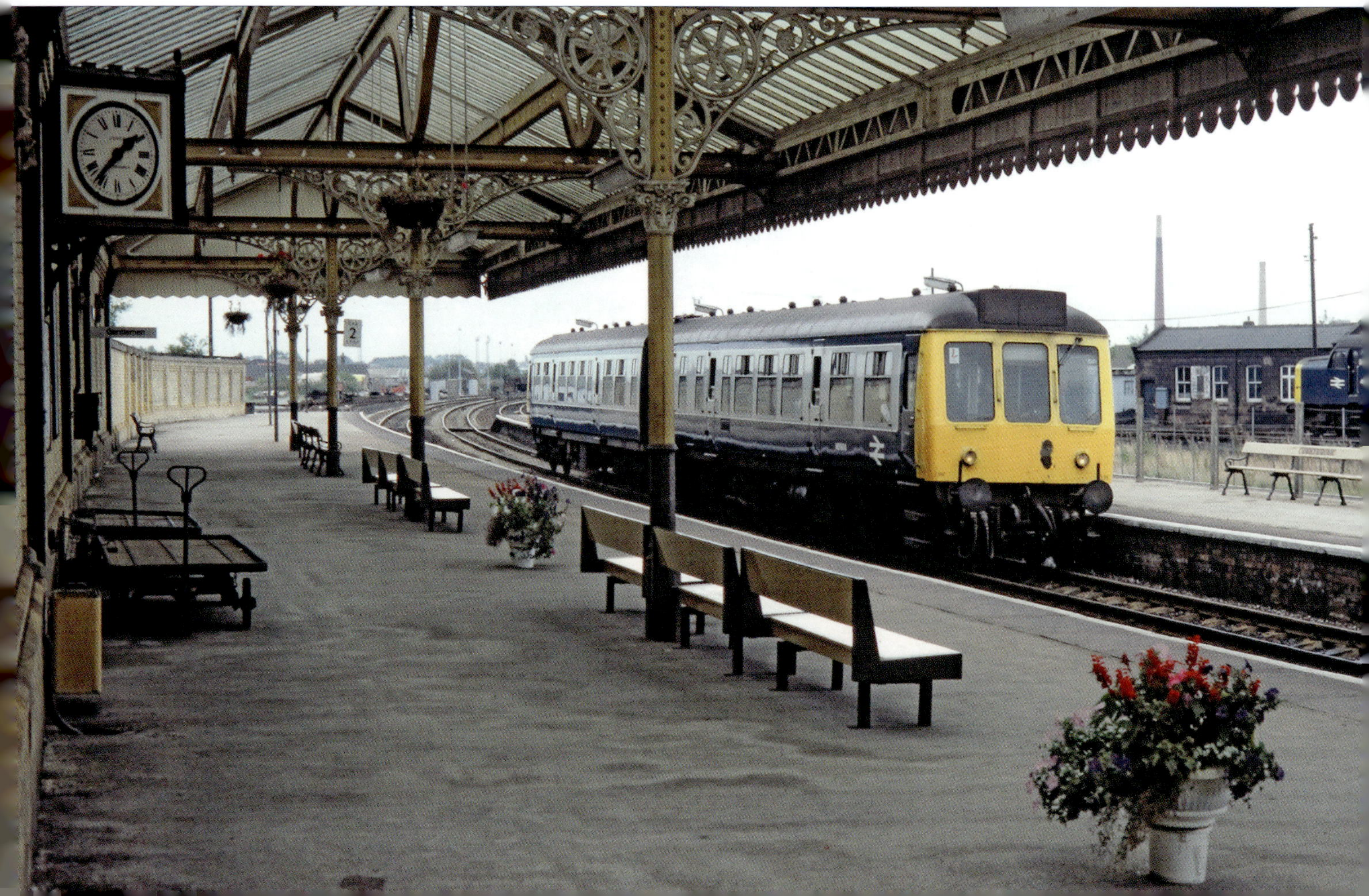
2

Top left. In 1875 the CLC opened its Northgate station in Chester, a modest structure which closed in 1969. But even getting to Chester was low priority for the joint railway. The Cheshire Midland Railway had arrived in Northwich in 1863, and by 1867 the newly-constituted CLC had opened a branch connecting to the salt mines and works around Wincham. Even before that there was parliamentary approval in 1861 for a link from Northwich to Helsby which was built in 1869 as the line from Mouldsworth (see also p.72) giving access to Birkenhead. Remarkably there was then no progress west of Mouldsworth towards Chester until 1874, and the survival until 1992 of the line as far as Dee Marsh Jct after the closure of Northgate station indicated the pre-eminence of freight over passenger traffic to the CLC and to BR. My first visit to Northwich shed, conveniently alongside the yellow brick station (note the rear view of the station turret clock) was on 27th February 1966, when the sad display of rusting Stanier and Ivatt locos also included recently-withdrawn and dismantled BR Standard class 3 2-6-0 no 77011. This was the smallest class of Standards with just 20 produced in 1954, all allocated with equal shares to the North-Eastern and Scottish regions until 1965 when nos.77011 and 77014 were despatched to Northwich (and no. 77014 on to Guildford in April 1966). Often regarded as either too big or too small for potential employment, what use (if any) did Northwich make of them? *HF-S141*

Bottom left On 14th September 1983 class 108 DMU no. 51904 is seen drawing into Northwich station with the 13.00 Manchester Oxford Road – Chester service. By then diesel railcars had long since taken over from superannuated mainly ex-GC steam that took up space in our Eastern Region locospotters ABCs but were never seen in Eastern Region territory. (The glimpse of EE type 4 no. 40001 on the right, formerly no. D201 which took the place of our 'GE' Britannias in 1958, did not compensate for the failure to see the 'Director' 4-4-0s.) The closure of Chester Northgate and diversion of services to Chester Central brought to an end the Mickle Trafford saga (1901 population 348) which from 1889 had two stations almost facing each other on unconnected lines. Northwich station, in the midst of so much industrial activity (note the tall chimneys of Brunner Mond's Lostock Works on the right), made a good effort to welcome passengers, from the elegant iron platform awnings and the handsome clock from Joyce of Whitchurch, to the station cat (unfortunately not present this day). The hanging baskets, in decline in September, were added to with blooms from platform pots. *IMG-0530*

Above. In 1867 the LNWR opened a single track branch line from Middlewich to Northwich. At Northwich the branch connected with the CLC line at almost 90 degrees, with the connection made by a spur to the east. Although a corresponding spur to the west was evidently planned it was not in fact built until 1959, by BR. In 1870 the LNWR had built a connection from Hartford Jct on the ex-GJR to the CLC to reach Winnington. Intentionally or otherwise, the 1959 spur opened a diversionary route from the WCML between Hartford Jct. and Crewe via Northwich, Middlewich and Sandbach, and on 11th June 1983 two trains from Liverpool Lime St to Euston (14.25 and 15.50 departures) took this route. With no information (then or since) as to the reason for the diversions or even how I learned of them, it is hard to see why only Liverpool departures (as it appears) would have been affected. Perhaps there were more, but being unfamiliar with the Northwich-Middlewich branch I had hoped to photograph something special. In fact it runs arrow-straight and the trains were almost brushing aside mid-summer luxuriant vegetation. Only the 15.50 with class 47 no. 47306 in charge with a pilotman also in the cab offered any interest when seen from Whatcroft Hall Lane bridge by the Trent and Mersey Canal, visible in the photo on the right before it bends sharply to pass under the line. *IMG-1719*

Merseyside and West Cheshire Railways 1965-1990

Below. I find industrial architecture fascinating and in its way beautiful, and this view looking south of the Brunner Mond Winnington Works to me is amazing with building seemingly heaped upon building. Winnington was located on the edge of Northwich, about a mile or so on a branch leaving the CLC line between Greenbank station and the viaduct over the Weaver and Dane Rivers. It was also accessed by a spur of the LNWR from its main line near Hartford to the CLC in 1870. Production of soda ash began in 1873, and the company was merged with others to form ICI in 1926. It was separated and sold to Tata (later Tata Chemicals Europe) in 2006 but the last trainload of limestone arrived in 2014 and production of soda ash and calcium chloride ended. The loco photographed on 14th September 1983 was *Perkin*, a diesel electric shunter built by English Electric in 1951 (works no.1904) and obviously akin to BR's class 08 (Perkin named for the 19th century industrial dyes research chemist). It is pictured on its way to collect seven yellowish ICI tank wagons. In the foreground is a stretch of the Weaver Navigation which about a mile to the left (east) connects with the Trent and Mersey Canal at Anderton Boat Lift. *IMG-0533*

Top right. To a first-time visitor to Crewe South shed in the evening of 27th March 1966 this extraordinary 60ft structure, like a skeleton with a head and a digestive system, should have been an eye-opener. But I probably dismissed it as having something to do with coaling before moving on to admire the two locos framing the picture, Standard Britannia class 4-6-2 no. 70032 (formerly *Tennyson*) and 'Black 5' 4-6-0 no. 45096. Nick Pigott's superb survey of 'Towers of Strength' reveals both the scale of the coaling problem at big sheds, and the historical significance of the Crewe South coaler, seen here. By 1910 Crewe North had installed a steel tower mechanical system based on coal wagons discharging their load into a ground-level bunker and a conveyor which lifted the coal nearly 50ft to be dropped into hoppers. With improvements it functioned until replaced in 1951. In 1912 Edge Hill adopted a 'staithe' system with coal wagons pushed onto an elevated ramp to disgorge into hoppers below, the significant innovation being reinforced (aka ferro-) concrete. Crewe South's contribution in 1920 was to pioneer a wholly mechanical system combining the huge strength of ferro-concrete with the capacity to lift coal wagons vertically 45ft on the steel tower and rotate them emptying into the ferro-concrete hopper which held 250 tons. Also with improvements, it saw out the steam age. Note the old-style grey mineral wagon on the platform about to be lifted. *HF-S161*

Bottom right. Class 3F 'Jinty' 0-6-0T no. 47530 is seen here at Crewe station on Platform 1A (the northern end of Platform 1) on the 1st October 1966, at 14.55 according to the platform clock. (Two days earlier N.E.Preedy photographed the loco in the station but without the '3' sign on the buffer plate, probably Crewe's designation of a particular set of station pilot's duties for this day.) Behind the loco are at least five parcels vans which it may have just drawn into the station. On the platform are four or five trolleys stacked high with parcels which suggests that they are waiting to be loaded rather than just arrived. The two on the right have cages with 'Down' chalked on them, and 'Bridle' on the one on the left (which must have meant something to somebody). No. 47530 was in fact withdrawn from service during the week ending 8th October so this may have been its final appearance in public, but other 'Jinties' rather than class 08 diesel shunters continued to have a role (if only briefly) with station pilot duties at Crewe. *HF-S1337*

Merseyside and West Cheshire Railways 1965-1990

5 WCML

C.M. & E.E.
LONDON MIDLAND REGION
AREA MAINTENANCE
ENGINEER
WIGAN.

CREWE STATION A

Top left. In early 1984 BR gave notice that Crewe station would be almost completely closed for seven weeks in midsummer 1985 to allow for a major programme of track remodelling and signalling. This restructuring extended to the approaches from north and south, the number and utilisation of station platforms as well as better facilities for passengers. The headline improvements would be raising the permitted speed of passing trains from 20-30mph to 80mph, two-way signalling on all through lines, and simplification of trackwork. 'A's and 'B's for bay platforms were abandoned, as was platform 1 entirely. Old 2 became new 12 for parcels, and only old 5 kept its number. Three were extended by up to 375ft. The budget for this project was £14.3m. I was pleased on 4th July 1985 to find there were no barriers to access at the station, and there was plenty going on to gaze at. The biggest surprise was the informal way the engineer worked on the overhead structures, with a ladder propped against class 97 (ex-40) no. 97408, one leg dangling, head protection and hi-vis waistcoat discarded on a hot day. (Subsequent frames show the engineer spread-eagled, hand and foot on the ladder, the other foot resting on a V-shaped angle in the structure, the other hand fixing something. Necessary and laudable, no doubt, but Health and Safety At Work Act 1974?) *IMG-0710*

Bottom left. Crewe Station A signal box was installed in 1907, an LNWR type 5 box which was accommodated with a flattened roof under the station canopy, between old platform 2 and the adjacent through line. On 4th July 1985 it had been relieved of signalling duties since 2nd June with the closure of the station for its remodelling, but was still required for setting points for any movements in the station. In this view class 25 no. 25095 (coupled with an invisible no. 25198) is seen on a crossover heading towards the platform while engaged on construction work for the station project. Crewe station was completely closed for seven weeks until 21st July 1985 with diversions arranged by buses or to alternative stations, but with two exceptions: one was a shuttle DMU service between Crewe and Chester, the other also a DMU shuttle between Crewe and Stafford. The photo shows a class 101 DMU with two passengers who have just left it, their conversation breaking the strange silence of a large station with almost no trains. Station A signal box was carefully dismantled at the end of the project and re-erected at Crewe Heritage Centre in 1987. *IMG-0714*

Above. On the 18th July 1967 Stanier 8F class 2-8-0 no. 48699 was passing over the magnificent Dutton Viaduct (between Acton Bridge and Weaver Junction) with a down freight of about twenty very mixed vehicles. The viaduct was built in 1836 by the GCR to take the line over the valley of the River Weaver, and was opened for traffic the following year. Built in local red sandstone, its design is generally attributed to the company's engineer George Stephenson, with Joseph Locke responsible for supervising its construction. It is comforting that a contemporary source says there were no recorded deaths or injuries, but the operative word must be 'recorded'. Its length has been variously stated as 428 or 500 yards but a quarter of a mile gives a better idea of it. There are 20 fine arches, half of them in the photo, each with 60ft span, the pillars having splayed bases which emphasise their massive strength, The loco is about to cross the main watercourse which since the 1870s has been the Weaver Navigation. What remains of the River Weaver's original course ran in front of the derelict 'Viaduct Cottages' (on the right) which have now disappeared. Water in the foreground is one of a sequence of fishing ponds extending either side of the viaduct. They are what is left of Pickerings Cut canal, dug in the 1760s. *HF-S1870*

Merseyside and West Cheshire Railways 1965-1990

Merseyside and West Cheshire Railways 1965-1990

Top left. In May 1962 EE/VF produced a prototype Co-Co diesel electric locomotive DP2 believing it would meet BR's need for additional type 4 traction. The same month VF also produced the last of the Deltics (D9021) and gave DP2 the same overall body shape. The interior was very different. The new EE 16CSVT power unit rated at 2,700bhp and its reliability over 600,000 miles on trials, coinciding with an increasing failure rate among early class 47s, persuaded BR to order fifty of class 50 (D400 series) for delivery in 1967-8. This in turn coincided with BR's decision in 1970 to electrify the rest of the WCML north from Weaver Jct to Glasgow (achieved in 1974). That left an interval requiring additional power to meet an accelerated timetable before the arrival of the thirty-five class 87 electrics, and that role was met principally by the 50s. What particularly attracted attention was the decision to adopt double-heading on the fastest / heaviest workings which accounted for 8 of the 34 diagrams on the line. The sound effects and ease with which coupled 50s dealt with Shap and Beattock were memorable to witnesses. No such dramatics were called for in crossing the Weaver valley. I revisited Dutton Viaduct in 1972 and watched a single 50 stride across the valley with an 11-coach load, and later (pictured) a pair doing the same. Soon after this encounter, the 50s were moving to the Western Region as deliveries of the class 87s came on stream. *HF-2022-31*

Bottom left. Standard class 9F 2-10-0 no.92224 was on the water troughs at Moore, heading north with a work train on 17th June 1967. Notice on the far right a large grey tower topped off with a cream coloured square structure showing some windows: more about this in HF-S1879. In the late 1850s John Ramsbottom, a Locomotive Superintendent of the LNWR, came under pressure to reduce the journey time of its trains, in particular the 'Irish Mail' to Holyhead. The challenge was to devise a way for locos to secure sufficient water to complete a long journey without needing to come to a halt. His solution was the water trough laid centrally between the rails, and a scoop lowered from the loco travelling at sufficient speed to force water into the loco's tender. Trials found the optimal loco speed for efficient water collection to be between 40 and 50 mph. Remarkably, by June 1860 the troughs were in use and during the decade the LNWR installed eleven sets, including this one at Moore. The LNW laid a further seven sets in the 1870s and 1880s, but other British companies were reluctant to follow suit. Only the L&Y in the 1880s (where Ramsbottom had become a director) and the GWR at the turn of the century made extensive use of water troughs. *HF-S1667*

Below. Standard Britannia class 7MT 4-6-2 no. 70014 (formerly *Iron Duke*) is seen here on 21st July 1967 with a down parcels. The location is a short distance north of Moore water troughs, in the cutting that housed Moore station until its closure to the public in 1943. The unmissable feature of the photo is the grey tower, mysterious when seen from the front in the previous picture, but ugly and menacing from this angle. Not a great back-garden sight for the houses over the wall! Nick Pigott names the reservoirs of water for troughs (which the Moore structure undoubtedly is) as 'tankhouses' (though tankhouses are also water tanks converted to provide living space) rather than towers, but either will serve. Unfortunately there seems to be very little written (or pictured) about their history and structure. Gratefully (and crudely) borrowing NP's calculations, the Moore tower might hold 40,000 gallons to replenish the 6,000 gallons in each of the Moore troughs, half of which would be lost to a loco 'using'. Did every set of troughs have its own tower to meet demand on this scale? Was there a standard LNW/LMS/BR model for the tower? What controls and facilities were provided for staff? How did they access the square structure on the top? Too many questions, not enough answers! *HF-S1879*

Above. With exhaust, blowing off and leaking steam swirling in all directions on a proper March day (26th, 1966), Standard class 9F 2-10-0 no. 92162 is battling the elements and a short climb at 1 in 135 with a load of mineral hoppers. The loco is passing through Moore village which, with a population in 1950 of 500, could (almost) claim two stations 400 yards apart and on the same NE-SW axis. The first Moore station on the WCML was opened by the GJR in 1837 and closed in 1943 though services for railwaymen continued until 1952 and perhaps later. The second Moore station was on the BLCJR, was earlier renamed Daresbury, and closed in 1952. The reason for the incline confronting no. 92162 dates back to the 1890s and the construction of the MSC. The elevation required for shipping on the Canal meant the LNWR had to build deviations for both lines. Just visible to the left of the picture are the original tracks of the BLCJR disappearing into a cutting at Daresbury Jct. as the 9F takes on the deviation leading to Acton Grange Jct where it joins the WCML. The old BLCJR tracks remained in use as sidings until the closure of Daresbury Jct signal box in 1972. *HF-S280*

Top right. Stanier class 8F 2-8-0 no 48151 is seen here at Moore heading north towards Acton Grange Jtn on the WCML on 1st April 1967. Clearly an effort had been made to smarten up the loco, but why the special treatment? Perhaps it was to celebrate its very recent transfer from Edge Hill shed to Northwich. It's also worth noting that the loco was retrieved from Barry scrapyard for private preservation in 1975. It thus belatedly joined the ranks of the preserved like the previous steam on this track a few minutes earlier, ex-LNER 4-6-2 no. 4498 *Sir Nigel Gresley* following its restoration at Crewe works. The 8F was working hard on the 1 in 135 gradient with its train of soda ash hoppers, which is ironic as it was here that another such train broke in half on the night of 13th May 1966, releasing runaway hoppers. They smashed into the cab of class 40 diesel loco no. D322 which was hauling the 20.40 Euston to Stranraer Harbour night sleeper, killing both crew members. Indeed the site of the crash is very evident from the soda ash still scattered around it, nearly a year later. In the foreground is the track bed of the GJR, in a shallow but deepening cutting, which led to a tunnel. The old track was only briefly made use of for sidings by the LNWR before the rails were lifted. The deviation also meant that the original bridge over the GJR supporting Moss Lane needed a new and rather ugly bridge. *HF-S1484*

Bottom right. Hail and Farewell! In the week ending 15th January 1966 Stanier 2-6-0 no. 42953 was withdrawn from service from Springs Branch (Wigan) shed. Perhaps this was to meet a quota of withdrawals or due to some misdiagnosed fatal defect rather than inability to raise steam, as on 2nd April 1966 the Mogul was seen in action on the WCML approaching Moore. It was entrusted with collecting new class AL6 electric loco no. E3151 from English Electric's Vulcan Foundry and delivering it presumably to Crewe Works for its formal induction and entry into service three weeks later. No. 42953 was cut up two months after its brief excursion. I also photographed the ensemble going away, which revealed that at the rear no. E3151's route indicator was showing 5T86. The loco retained its 'E' number until BR's adoption of the TOPS numbering scheme, when in September 1973 it became class 86 no. 86212. At that stage it would have lost its cast aluminium 'lion over wheel' crest in favour of the 'double arrow' insignia. In 1979 it received the name *Preston Guild* which it retained with the addition of the years *1328-1992* until withdrawn from service in October 2003. Notice beyond the caravans the BLCJR (Chester to Warrington) line is converging with the WCML. *HF-2-45-20*

Merseyside and West Cheshire Railways 1965-1990

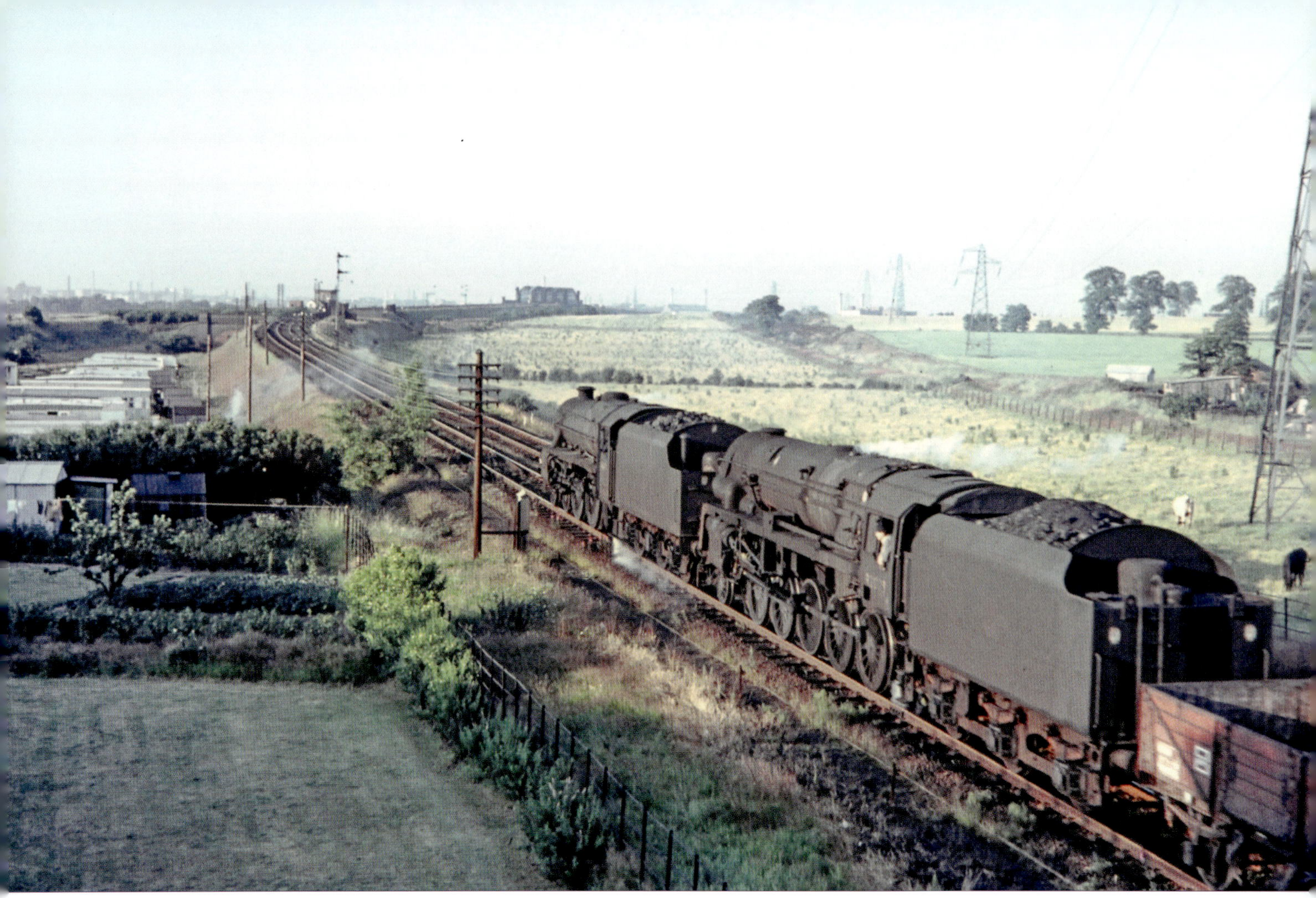

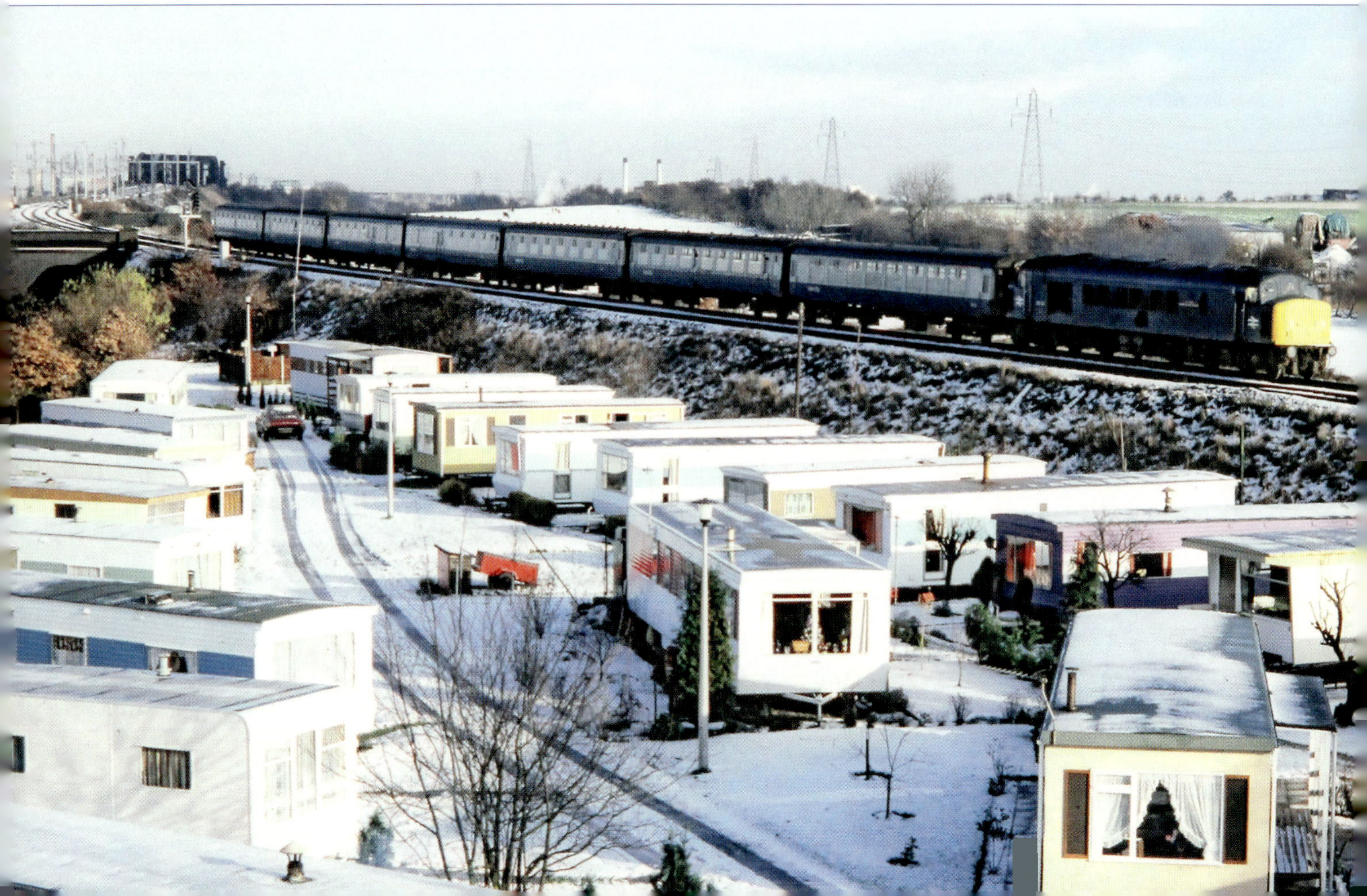

Top left. Opened in 1893, Acton Grange Viaduct is a massive steel girder structure which carries the railway 75ft over the waters of the Manchester Ship Canal. As well as the four lines each in a separate section of the bridge, its size is increased by the very oblique angle at which it crosses the Canal. It is visible as the large dark mass in the top of the picture which on 10th July 1967 has 'Black 5' class 4-6-0 no. 44897 piloting Standard 9F class 2-10-0 no. 92029 on the 1 in 135 climb to Acton Grange Jct and on to the Viaduct. Their burden is a train of Shell Oil tanks with a buffer wagon deployed between the loco and the tanks, as required by safety regulations. No. 92029 was the last of ten (92020-9) Standard 9Fs to be built as Franco-Crosti 2-10-0s in 1954-5. As a means of increasing fuel efficiency these 'Crostis' found room between the frames for a long drum beneath the boiler in which feedwater was pre-heated by exhaust gases on their way out. Unfortunately the anticipated efficiency gain was not achieved. To accommodate the drum the 'Crostis' had been fitted with a smaller boiler (BR9A) than the other 9Fs, and when they were rebuilt along conventional lines in 1961-2 they retained the BR9A effectively reducing their rating to 8F. The last eight (including 92029) were withdrawn in November 1967. *HF-S1817*

Bottom left. On 12th December 1983 'Peak' class diesel electric 1Co-Co1 no. 45104 *Royal Warwickshire Fusiliers* led the 08.50 cross-country service from Scarborough to Bangor on a snowy morning. The train has just passed Acton Grange Junction so it is now running on the deviated section of the BLCJR for another mile or so. Compared with the previous picture (HF-S1817) the flat-roofed Acton Grange signal box has disappeared after the opening of Warrington power box in 1972, and the semaphore signals are also gone. The foreground in the picture, a caravan park site, lies in the V-shaped valley created by the two converging deviated lines, but this wider view of the valley reveals a tunnel entrance at the valley's apex. The original GJR line northbound passed from a cutting into Acton Grange Tunnel whose exit would have disappeared for lack of use or in the MSC construction work. The BLCJR, the GJR's close neighbour to the east, arrived thirteen years later in 1850 and managed to find a route to Warrington without needing a tunnel. No. 45104 was built at Crewe in 1962 and was one of the last 'Peaks' to remain in service to 1988. The wheel configuration was shared only with the earlier class 40, both needing unpowered axles front and rear to keep within BR's axle-load limit of 20 tons.*IMG-0578*

Above. Between Acton Grange Viaduct (AGV) and Warrington the pattern of railways becomes complex. After less than a mile north from AGV the new line encountered the river Mersey, for which a new steel girder bridge ('Mersey Viaduct') was built. Like AGV it carried four lines but in 1968 the direct connection to Warrington Bank Quay station (the WCML) was reduced to two lines, with the eastern half of the 'MV' left to rust away. The gradient down to Bank Quay station exactly matched (1 in 135) the climb up from Moore to the MSC. On 10th July 1967 Standard 9F 2-10-0 no. 92046 is seen crossing the Mersey on its descent towards Warrington with Shell Oil tanks (and no barrier vehicle behind the loco: were safety regulations changing?). Daylight visible between the 9F's boiler and frames indicates why this was the only BR Standard class that could find room for the Franco-Crosti pre-heat system and even that needed the smaller BR9A boiler. Looking east beyond the WCML is another bridge crossing the Mersey, but at a lower level. This is the original route of the GJR from Bank Quay with its graceful 1837 red sandstone 'Twelve Arches' Viaduct, but having crossed the Mersey here traffic southward has either to manage a gradient of 1 in 69 to reach AGV over the Ship Canal or to remain north of the Canal in Walton Yard sidings. *HF-S1824*

Above. In the early 1970s BR decided that it needed a powerful specialist freight loco rated type 5 to handle heavy loads (in steam days the role of the 9F 2-10-0s). The choice of builder was restricted but Brush Traction designed a class 56 very similar in appearance and structure to its class 47. Brush sub-contracted production of the first 30 to Electroputere in Romania for delivery in 1977. The next 85 arrived from BREL Doncaster during 1977-83 and the final 20 from BREL Crewe in 1983-4. The Romanian batch suffered from multiple flaws that required immediate attention, but the class as a whole kept an unfortunate reputation for relatively high maintenance costs. On 30th July 1983 two were simultaneously with coal hopper trains at Walton just south of 'Twelve Arches' bridge, nos. 56093 (shown here, loaded) and 56006 (probably empties). No. 56093 was two years old and had recently been back in works receiving the full paint treatment including black buffers fringed white. To the left is the 'new' Mersey Viaduct with the nearer half no longer is use while engineers are at work on the far side. On the right are steps up to the 1837 'Twelve Arches' with the cooling tower of Thames Board Mills in the distance. No. 56006 was not a pretty picture, work stained in old BR blue with no embellishments, but remained active 26 years and is preserved by the East Lancashire Railway Class 56 Group. *IMG-1758*

Top right. Warrington Arpley station was a grand structure which opened in 1854 on the south-east side of the town and connected the SHCRC from Garston and Widnes to the Warrington and Stockport Railway (WSR). When Bank Quay station was re-positioned in 1868 the line to Arpley was served by low-level platforms under (new) Bank Quay for the line's passenger services, but Arpley station itself closed in 1958 and was demolished in 1968. However, 100 years earlier Arpley Jct had opened a route for the WSR to the south leading to the growth of freight infrastructure at Arpley Exchange Sidings north of the Mersey and Walton Yard south of the river. In this view (10th July 1967) of the Twelve Arches Viaduct 'Black 5' 4-6-0 no 45081 was heading south over the Mersey with a train of vans either for Walton Yard sidings or directly up the 1 in 69 to the MSC Viaduct. Were there normal service passenger workings over the Twelve Arches up until closure of the Latchford MSC crossing in 1985? On 31st July 1982 class 40 no. 40104 took at least eight well-populated coaches over Twelve Arches southbound which I guessed might be a summer Saturdays only service from the West Riding. *HF-S1830*

Bottom right. This is the view from Arpley looking south on a snowy 9th February 1985 with a quartet of class 25s (L to R nos. 206, 287, 245 and 051) on the move. Which way? My notes indicate they were approaching the camera but if so they were running northbound 'wrong line', with Arpley Jct. and the two-road shed (sub- to Dallam?) which closed in 1963 just round the corner to the left. If my notes are wrong and the 25s are southbound, they have just passed Arpley Grid Iron Jct. North (the double track in the bottom right corner) and are thus bypassing the fifteen or so tracks of Arpley Exchange Sidings probably en route to the low-level Mersey crossing at Twelve Arches. Over 300 of the 25s were built during 1961-7 in two distinct body shapes: nos. 245 and 051 were built with gangway front doors for multiple working (hence with shortened central windows) and multiple body-side ventilation grilles, nos. 206 and 287 have full central windows and cantrail level grilles. The huge factory with chimney and cooling tower on Arpley Meadows behind the locos was the works of Thames Board Mills, built in 1936 to produce paper and cardboard and closed in 1983, two years before the photo. *IMG-1041*

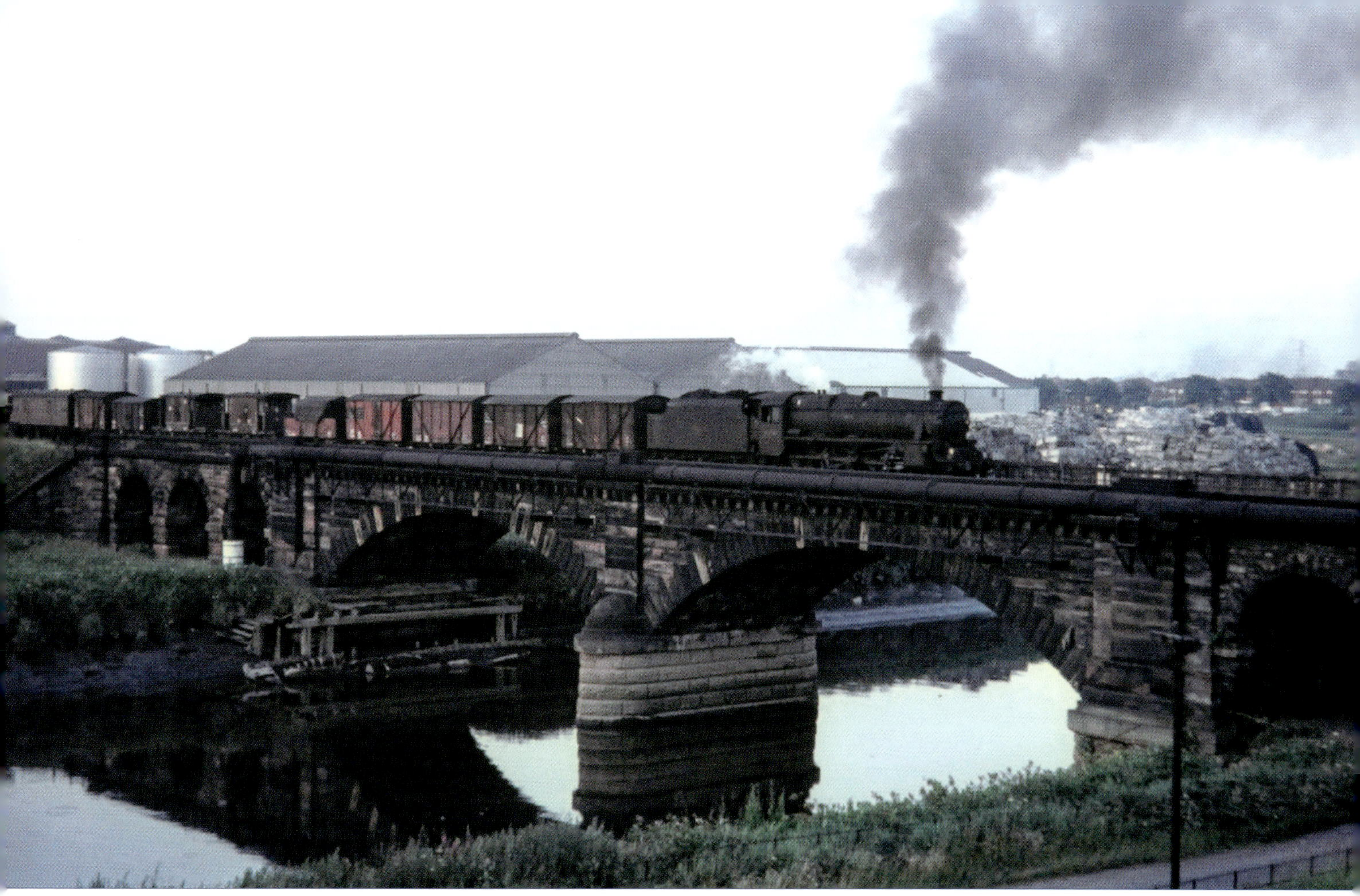

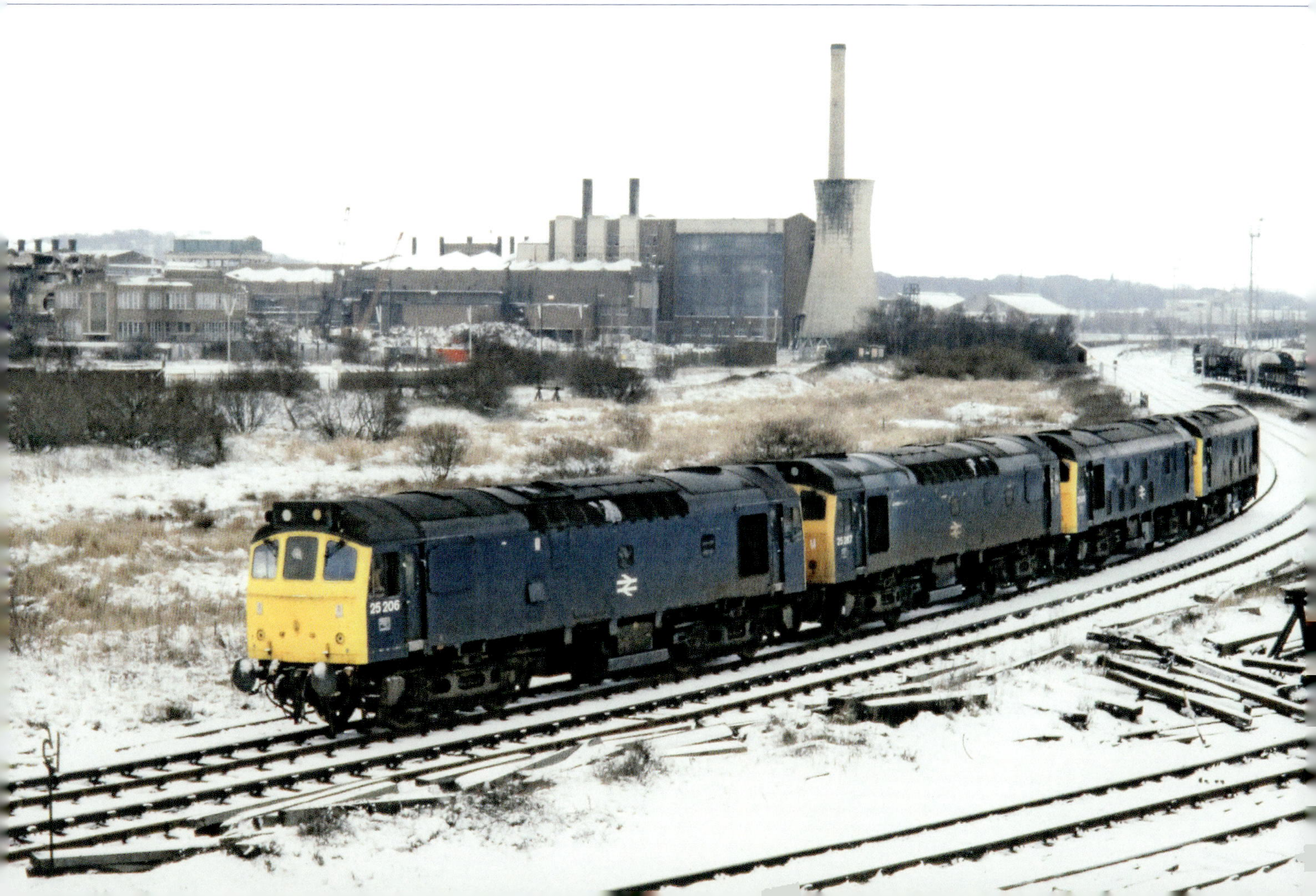
25206

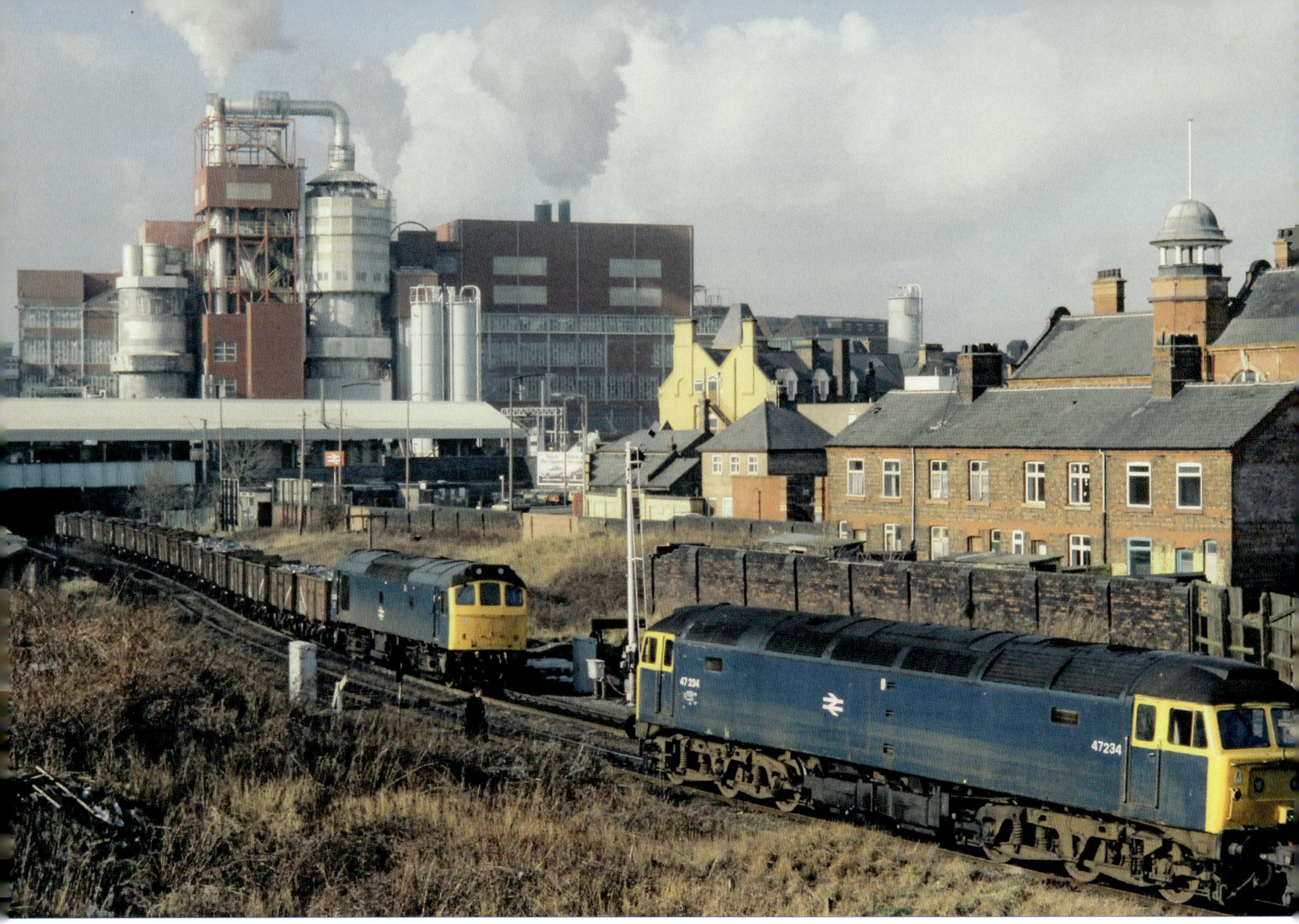
47234
47234

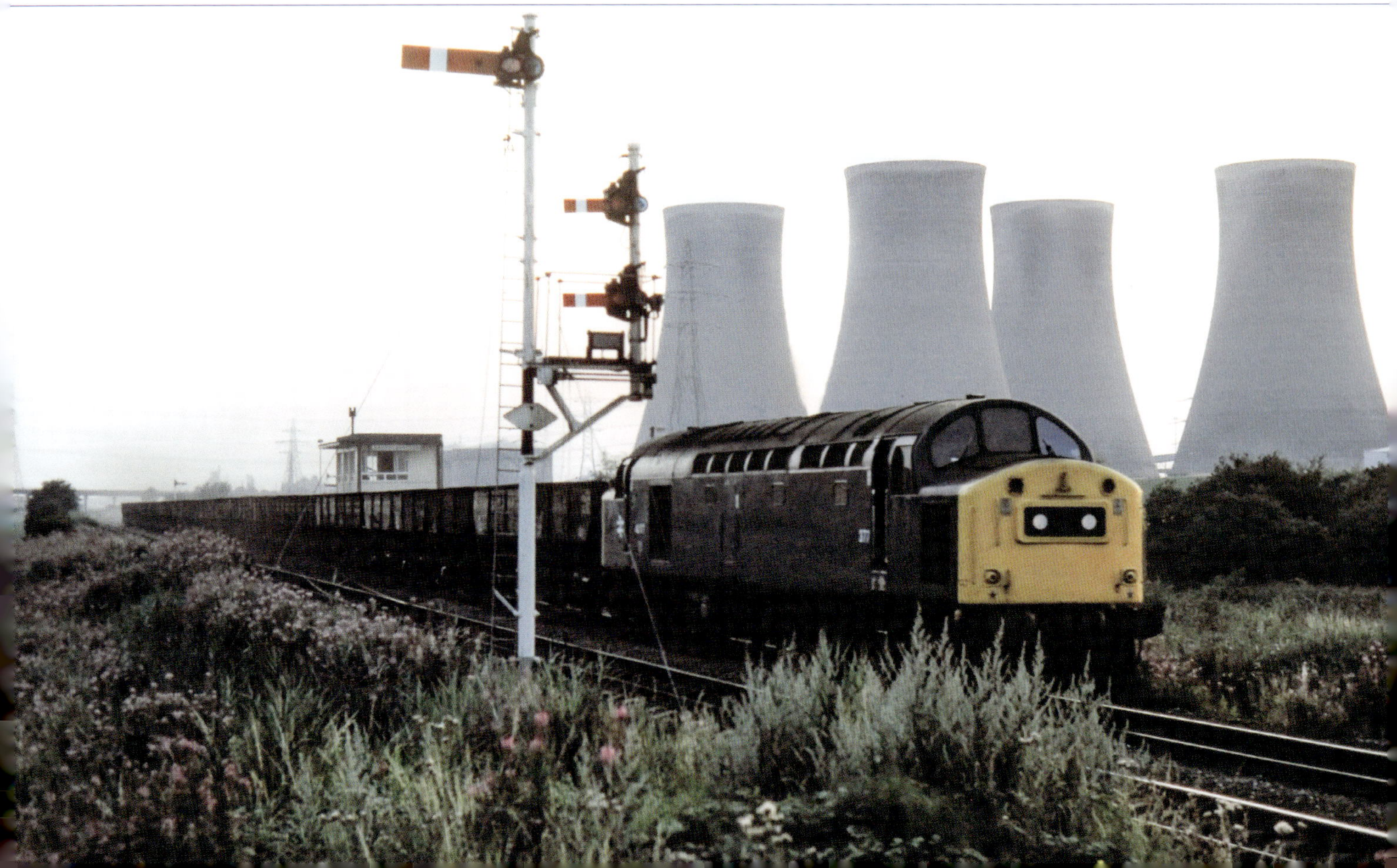

Top left. Arpley Jct was (and still is) well served by a convenient road bridge for observing the railway, and the view here looking west is of the line passing below Bank Quay station. Passenger services from the low-level platforms had ended by 1962 but the line carried a heavy volume of freight and Arpley shed closed to became a busy stabling point and service depot. On 9th February 1984 class 47 no. 47234 was approaching Bank Quay from the east with a train of the basic HAA hoppers (the first just outside the picture numbered 355-883) carrying coal for Fiddlers Ferry power station. The station was built with a Merry-Go-Round (MGR) loop for delivery of the coal, which in turn required locos to have Slow Speed Control gear which most of class 47/0 (including no. 47234) had. Heading east was class 25/2 no. 25279 which was emerging from under the WCML with an unidentified cargo (my notes just say '?wire and aluminium'). Dominating the scene from the other side of Bank Quay is the Unilever soap and detergent works which also closed in 2020. The Northern Powerhouse Rail proposals of 2014 envisaged a transformed role for the route between Liverpool and Manchester via Ditton Jct and Bank Quay low level, but nothing is currently in prospect. *IMG-1331*

Bottom left. Building Fiddlers Ferry power station began in 1967 for the state-owned CEGB and opened in 1971 before privatisation in 1991. Its eight cooling towers and 650-ft chimney dominated the landscape and was visible from far and wide. The station also burned biofuels but it was basically reliant on up to ten daily deliveries of coal (or about 15,000 tons) by rail when in use to capacity. Locos mostly of classes 20, 47 and 56 were fitted (or built) with electronic Slow Speed Control which was required by the MGR system for continuous discharge of the fuel. However, the SSC-enabled locos did not include the class 40. One of the surviving 40s, no. 40177, is seen here passing Fiddlers Ferry eastbound with a train of mineral hoppers on 29th July 1982, but the length of the train (stretching beyond the junction) as well as the lack of SSC rules out a Fiddlers Ferry connection. Rail operations at the station were complicated by exclusive access to the circular loop from the Warrington (eastern) end of the Warrington-Widnes line, a short distance beyond the wooden flat-roofed signal box in the photo. Coal deliveries were initially and conveniently from the Yorkshire coalfield to the east, but after the miners' strike in 1983-4 followed by pit closures much of the coal was imported through Birkenhead and Liverpool inconveniently arriving from the west requiring at least one reversal. *IMG-0877*

Above. On 27th September 1986 Fiddlers Ferry power station threw open its doors to welcome visitors to an Open Day. Class 20 Bo-Bo diesel electric no. 20157 (one of the 1966 batch from Vulcan Foundry) and its class-mate pair were handily parked with their train of hoppers by a building entrance, and attracted a lot of attention. Observed through a rather greasy window, folk of all ages queued patiently for their turn in the cab, an opportunity they had probably not anticipated when dressing in best leisure clothes for the occasion. At the time I had failed to appreciate the significance of the class 20s which were increasingly employed at Fiddlers Ferry in the mid-1980s. Toton had a large allocation of 20s equipped with SSC, and a dozen pairs of these were outstabled at Wigan Springs Branch depot. They were employed on hopper trains from Bickershaw Colliery to Fiddlers Ferry, a journey involving three reversals. I learn that the 20s were capable of operating at full power while at slow speed, whereas the 47s were susceptible to damage when operated in this way. Moreover the 20s could handle trains of 45 HAA wagons and were less prone than larger single locos to slip owing to their gearing ratios. *IMG-1570*

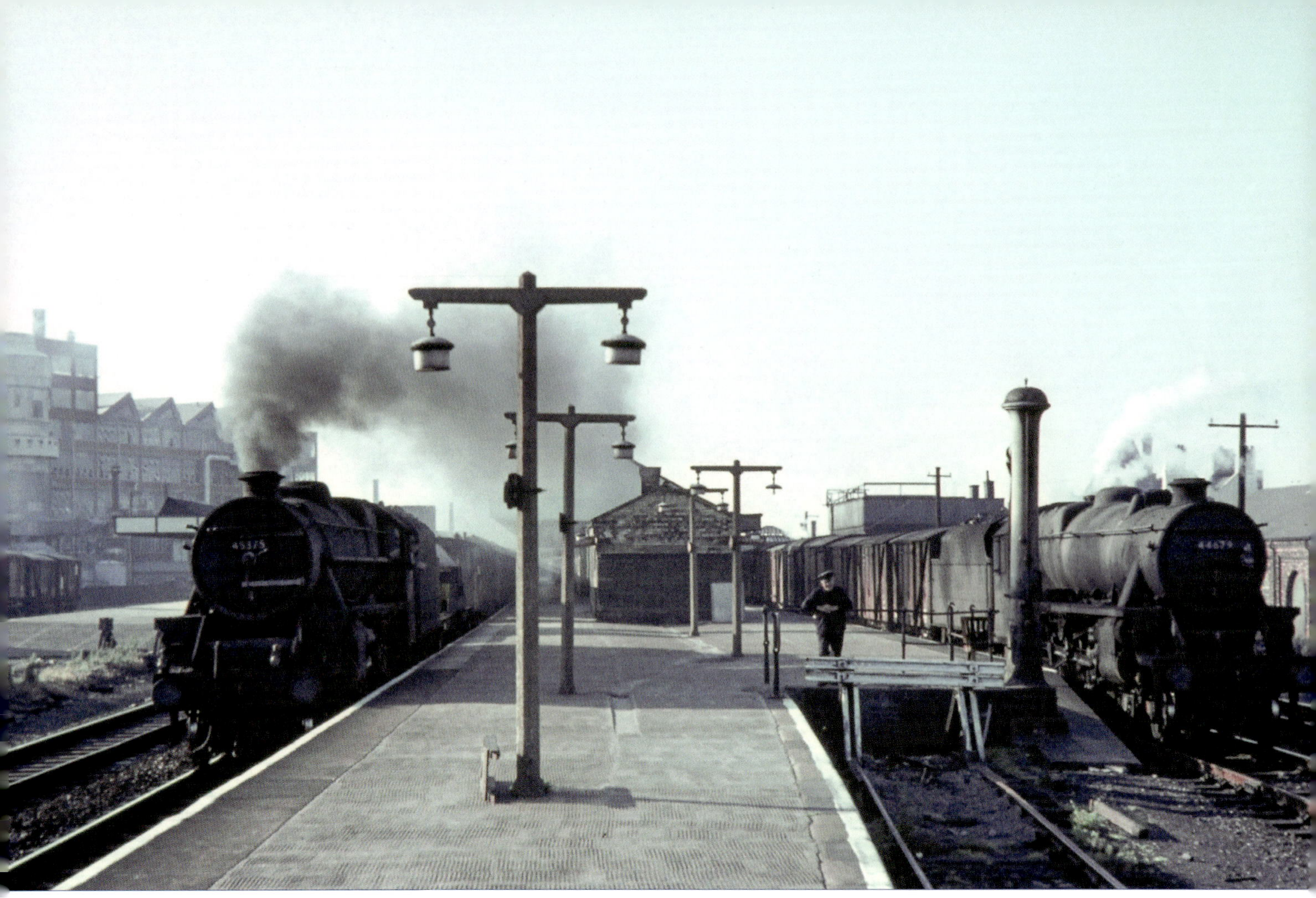

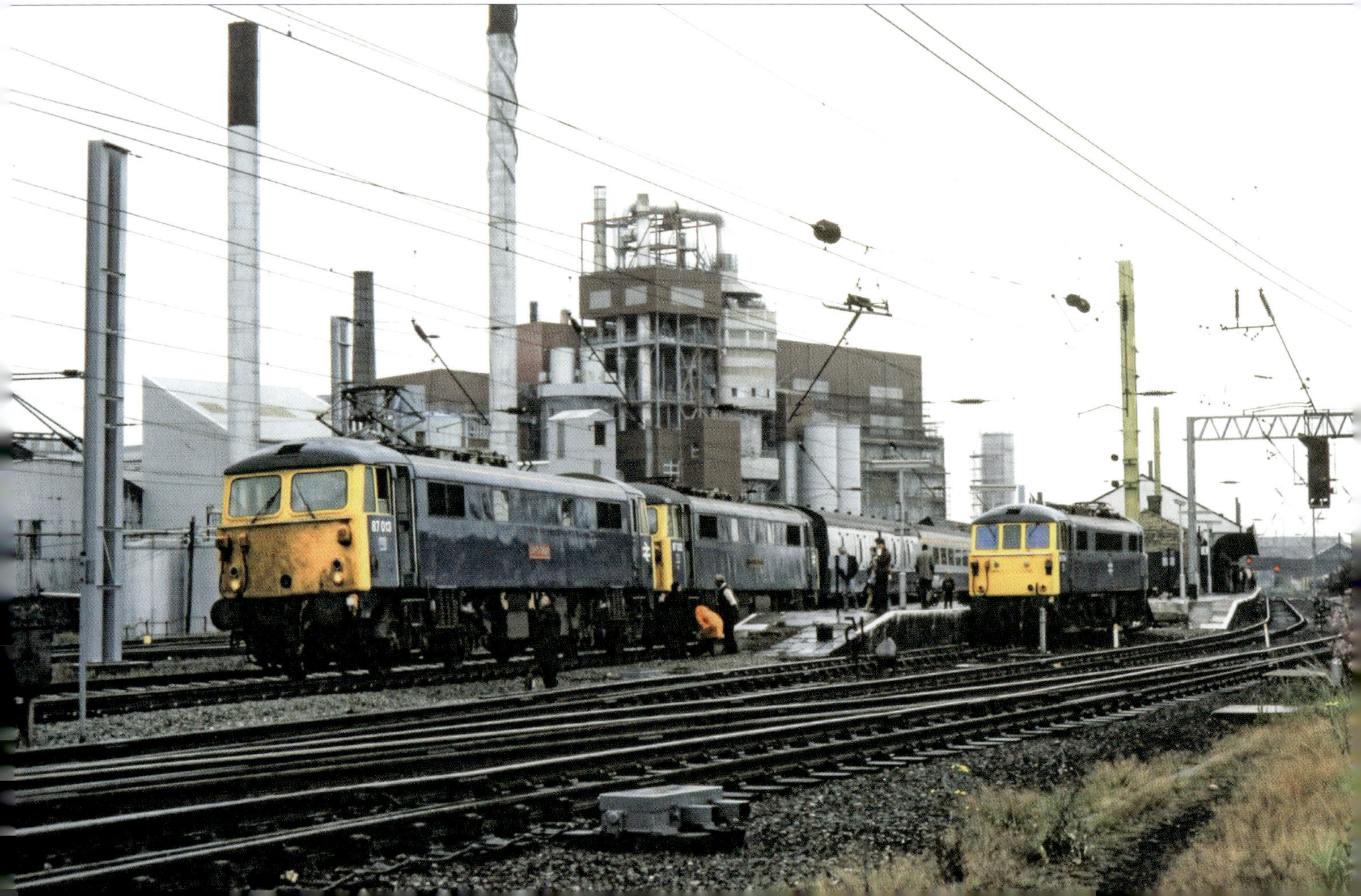

Merseyside and West Cheshire Railways 1965-1990

Top left. Warrington Bank Quay station on the warm afternoon of 10th July 1967 was as usual redolent with the sweet smells of cleansing agents. Waiting on the slow line at the south end was 'Black 5' 4-6-0 no. 44679 with a train of vans. Its driver had left the footplate and was leaning on the railings waiting to get the road. The peaceful scene was interrupted by another 'Black 5' no. 45375 which raced through the station with another fast freight, leaving a trail of smoke. The photo confirms what was just a fleeting view, that the leading vehicle in the train (no barrier) was a spent nuclear fuel flask from Cumbria being conveyed on a four or six axle 'Flatrol MJ'. At that time such flasks were easy to recognise by their shape (a white flattened conical cap) and colour (dull yellow), but they appear to have had no name other than to describe what they were. The flasks seem to have been invariably marshalled next to the loco in a mixed freight. Shortly after the disappearance of no. 45375 it was followed on the up line by no. 44679 with its vans. Meanwhile two 9F 2-10-0s (an ex-Crosti as pilot) were visible from Bank Quay station with a Shell oil tank train (two wagons as barriers) leaning into the curve approaching Arpley on their way to the West Riding. *HF-S1812*

Bottom left. Apart from the pairs of class 50 diesel electrics in the early 1970s, double-headed trains with two of the same class were not that commonly seen at Warrington Bank Quay station. But the morning of Saturday 31st July 1982 produced two of such trains, in rapid succession. First up was an incident affecting the 07.10 Glasgow-Euston service with train engine class 87 Bo-Bo electric no. 87022 *Cock o' the North* which evidently had a problem, causing stable-mate no. 87013 *John O' Gaunt* to be attached as pilot. That at least seemed to be the plan, but there were three wise men in discussion where the locos were being connected, and another calling up to the driver of the pilot. Class 86 electric no. 86004 was on stand-by duty in the bay platform, but not needed. I then left the station and walked round to Arpley, just in time to witness the departure of the 07.56 Leeds-Llandudno with a pair of class 40s in command, with no. 40030 (formerly *Scythia*) piloting no. 40087. Unfortunately lack of space precludes showing the picture here, but judging by the number of heads out of windows there were a lot of happy passengers in the front carriage. *IMG-0880*

Above. Warrington Dallam loco shed, was built in 1888 by the LNWR a mile north of Bank Quay station alongside the WCML. When viewed from Folly Lane on 23rd June 1965, Dallam shed forecourt offered a good variety of types all in steam including Jubilee class 4-6-0 no. 45563 (formerly *Australia*) next to a Jinty 0-6-0T on the right, and in the centre a Standard class 9F 2-10-0 partially obscuring Fowler (LMS) class 4F 0-6-0 no. 44115. Also present but wholly obscured were another 9F 2-10-0 no. 92124 and 'Black 5' 4-6-0 nos. 44658 and 45442. From the extreme left of the picture is a structure running to the right which disappears from view behind drifting steam and the loco shed. This elevated structure about two miles long was part embankment, part bridges (three in the photo) and another (behind the shed) over the WCML, and dates from 1883. The rest of the CLC's east-west main line from Trafford Park to Cressington was built in 1873 and included a loop to the south: local pressure in Warrington obliged the Committee to interrupt their intended straight line by placing Warrington Central station closer to the middle of the town. The CLC completed the elevated straight route ten years later, closing the gap between Padgate Jct and Sankey Jct thus enabling freight (until closure in 1968) to avoid passing through Warrington Central. *HF-1-11-12*

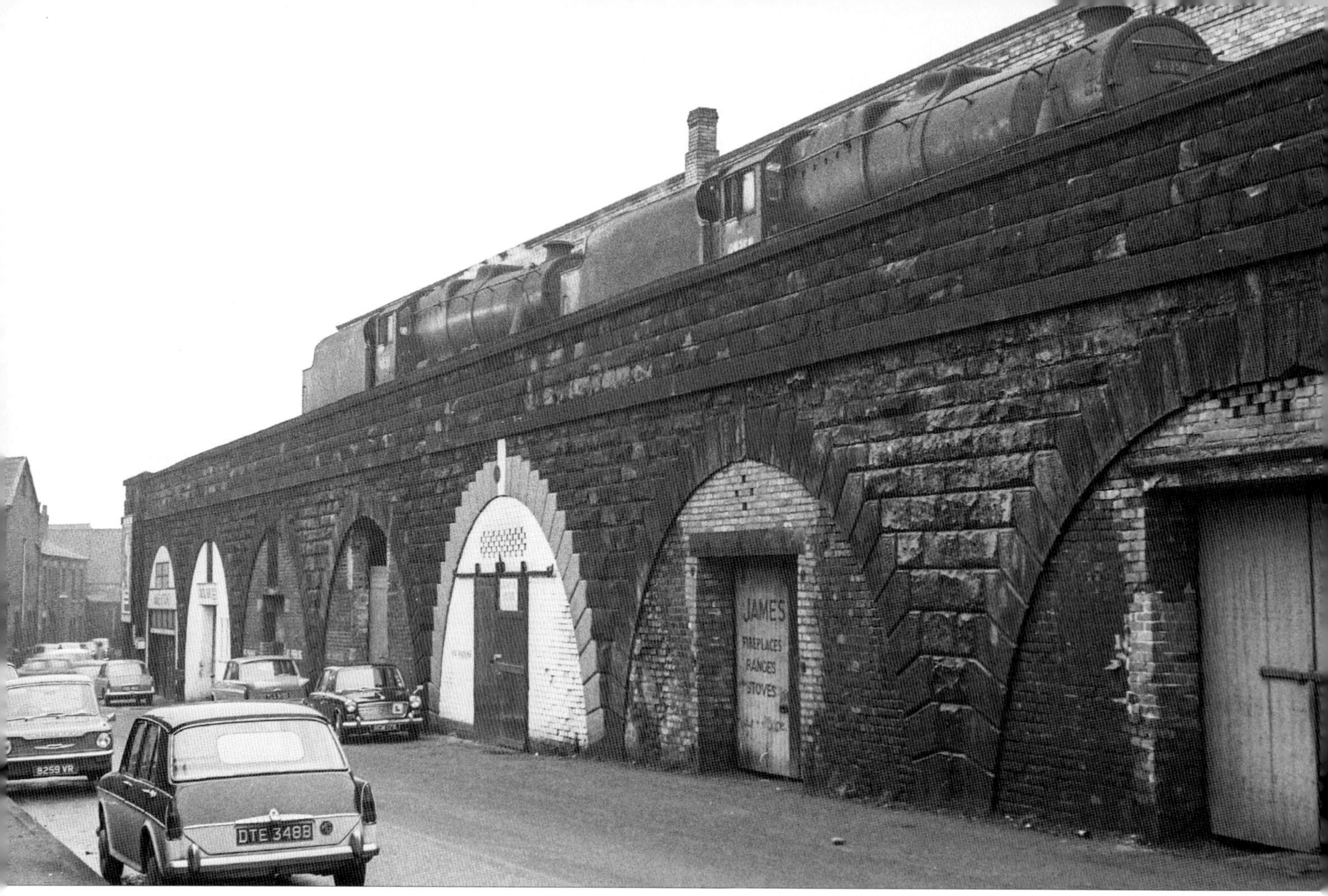
JAMES
FIREPLACES
RANGES
STOVES
DTE 348B
8259 VR

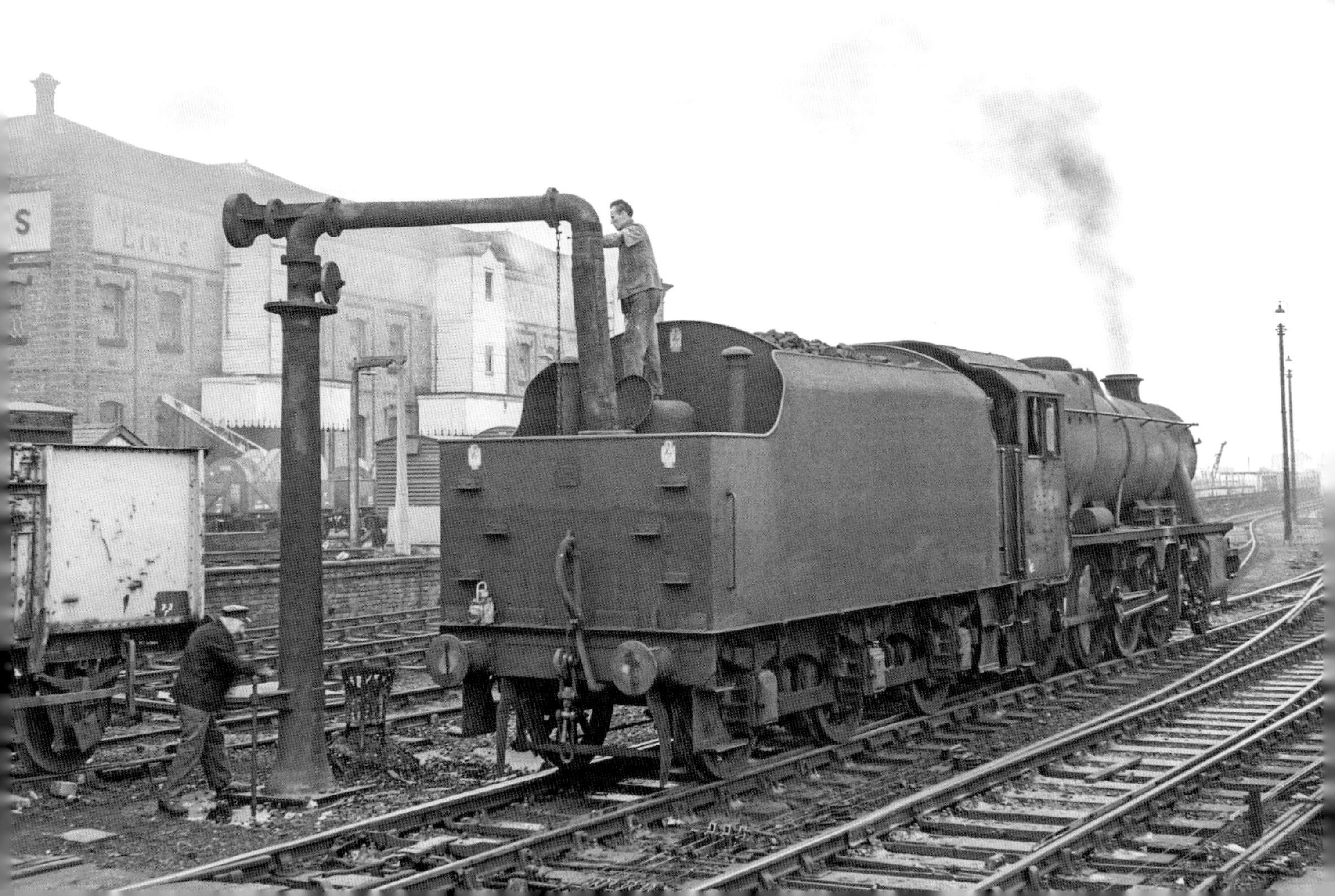

Top left. The CLC's Warrington Central station was built about a thousand yards closer to the town centre than planned on an elevated site above street level. The main building, long and low, is a fine structure but it faced north, away from the town leaving its aesthetic quality difficult to appreciate. The flank of the station readily visible to townspeople was overlooking Crown Street (now Midland Way) and was sturdily built with blocks of stone and handsome arches giving access to non-railway storage areas. This rear view of Crown Street was photographed on 28th January 1967 looking a bit of a backwater though obviously useful as somewhere to leave the car. But it is startling to see in view two class 8F 2-8-0s nos. 48356 (in steam) and 48350 (nearer the camera). This was 'Warrington CLC' loco shed, formerly a sub-shed of Liverpool Brunswick which had closed in 1961. In recent times the strip of land which accommodated the locos between the back of the station and Crown Street now fronts onto Midland Way. Removing the stone wall has revealed in the top right corner of the photo the original pale brick wall of the station. *HF-4-71-20*

Bottom left. Unfortunately I did not speak with them and have no idea who these two railwaymen were (photographed on 28th January 1967) but I can only speculate that the driver would probably be ending his career soon, when the skills he had acquired over a lifetime would soon count for little in the railway industry. Whether or not he is offered retraining, the prospect is of an uncertain, less secure future. The young fireman, aloft on the tender in his traditional role of putting the bag in to take on water, might question why he finds himself still having to work with an old and outdated technology. Why bother? But beyond the different perspectives of different generations the team nonetheless gets on with the job, out of pride striving to get the best out of their loco, class 8F 2-8-0 no. 48650, in its last nine months before scrapping. In the background of this scene at the eastern end of Warrington Central station is the huge and very impressive 1897 goods warehouse with its intricate brickwork and the names of the constituent companies of the Cheshire Lines writ very large on concrete panels. Sadly, listing Grade II of the three-storey building did not protect the three hoists, two of which appear in the photo. *HF-4-72-23*

Above. The country stations on the CLC's main line between Liverpool and Manchester were designed to look welcoming rather than to impress. The basic pattern linked a two-storey building with a similar but single-storey, the space between them shielded by a sloping roof resting on four columns on a recessed area of platform. The effect was to create a snug waiting area with a bench seat provided for the further comfort of waiting passengers. Padgate station, seen on 8th December 1985, had recently been repainted, but had unfortunately lost the very decorative wooden bargeboards to the gable ends (though it retained them on the small building [not shown] on the opposite platform). Track relaying was in progress, with 08 class shunter no. 08524 in charge of the non-self propelled machine seen here with its two jibs deployed as a new section is lowered into place. Royston Morris informs us that 24 machines were built between 1952 and 1975 for track relaying on BR, but the technology must have advanced massively in that time. Is this machine in fact no. DRB 78123 illustrated in RM's book? Also in attendance were departmental loco no. 97406 (ex-40135) behind the mess vehicle, and class 25 no. 25282. *IMG-1228*

Above. In 1831 the Warrington and Newton Railway built a line of four miles from Newton Jct (later named Earlestown) on the LMR to Warrington, anticipating the connection with the GJR which was finalised in 1837. For much of the WNR's route it was accompanied by the St Helens Canal (visible in the photo as a pale line on the left), nowhere closer than at Winwick Quay where a boat repair yard with workshops, moorings and a dry dock had been established. This may have encouraged the WNR to build Winwick Quay station, but there was probably no trade between railway and canal and the station closed in 1840. But the name lived on at Winwick Quay signal box, which on 20th May 1966 saw Standard class 9F 2-10-0 no. 92090 passing on the up slow line with a mixed freight. The importance of the box was the dozen or so sidings behind it on the east side of the WCML which was quadrupled in 1881 as far as Winwick Jct. Winwick Quay itself is visible as distant buildings below the down signal gantry: the boat repair building is end-on between canal and railway, with the 'Ship Inn' just visible at its west end. The M62 now passes just to the far side of these buildings. *HF-S466*

Top right. On 13th May 1967 class 8F 2-8-0 no. 48460 has just passed the Winwick Jct signal gantry at the head of a long mixed freight with a banking loco 'Black 5' 4-6-0 no. 45303 giving assistance at the rear. The freight is leaving the WCML and heading north to Earlestown on the WNR route then taking either of the tightly curved West or East Jcts onto the LMR. In 1864 the LNWR created Winwick Jct with an important link for the future WCML. The link by-passed Earlestown and burrowed under the LMR to join the Wigan Branch Railway at Golborne where the 1881 quadrupled main line from Warrington, interrupted at Winwick Jct, was restored to four tracks in 1888. This freight train was the only instance of banking that I witnessed here, but the climb from Winwick to Earlestown included a mile at 1 in 88 at Vulcan Halt so it must have occurred quite often. How would the decision be taken whether to request a banker? It was the guard's job to ensure the safety of the train (it was the guard who was held responsible for runaway wagons in the Moore crash in 1966), so he would report to the driver having inspected and estimated the load, and presumably the driver might then ask for assistance. In this case it is likely that the banker was provided by Dallam shed. *HF-S1536*

Bottom right. Brought up as I was on the side of the country where boilers were generally parallel and fireboxes round-topped, it was a happy surprise on 10th December 1966 to see K1 class 2-6-0 no. 62065 passing Winwick from Earlestown via East Jct with a mixed goods. The K1s were built by the Glasgow company North British Locomotive (NBL) for BR in 1949-50, but the design had deep roots in the LNER. One of Gresley's 3-cylinder 2-6-0s of 1937-8 was rebuilt by his successor Edward Thompson in 1945, and the externally unchanged design was taken forward by the LNER's last CME Arthur Peppercorn for production of 70 units for BR. The ER's share was 30, all initially allocated to March (31B). The rest were distributed around the North Eastern Region though none went to York at first, but 50A had become no. 62065's home shed when by chance I had seen it there four months previously. WDs (which for wartime austerity also defied the tapered boiler / Belpaire firebox norm) frequently worked through from Yorkshire, but the much smaller no. 62065 seemed to still have plenty of coal in the tender. Where did its journey start and end? I've found no record of this working. Notice VULCAN spelled out on one of the shops at Vulcan Foundry above the signal box. A few months earlier the complete legend with LOCOMOTIVES had been there. *HF-S1361*

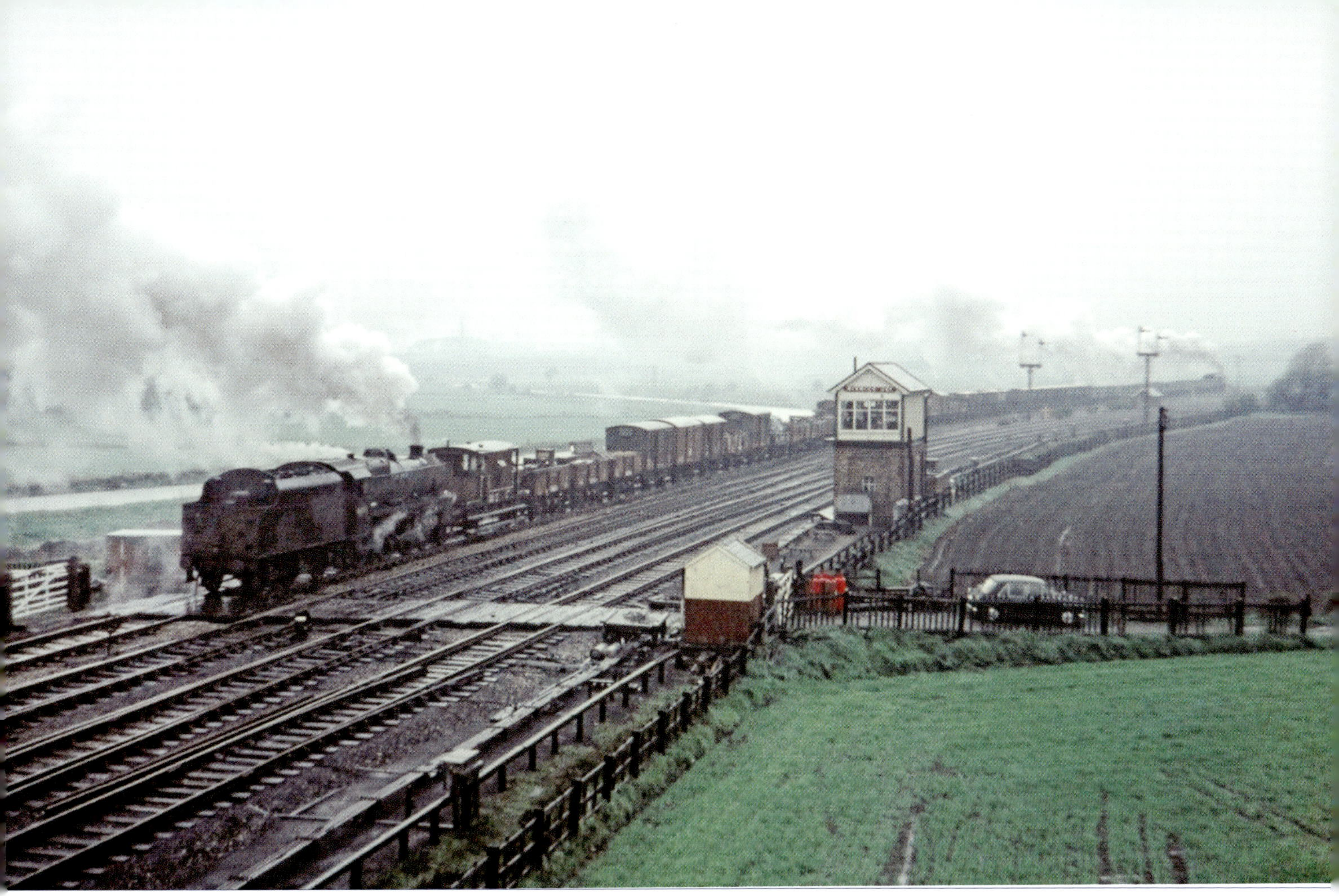

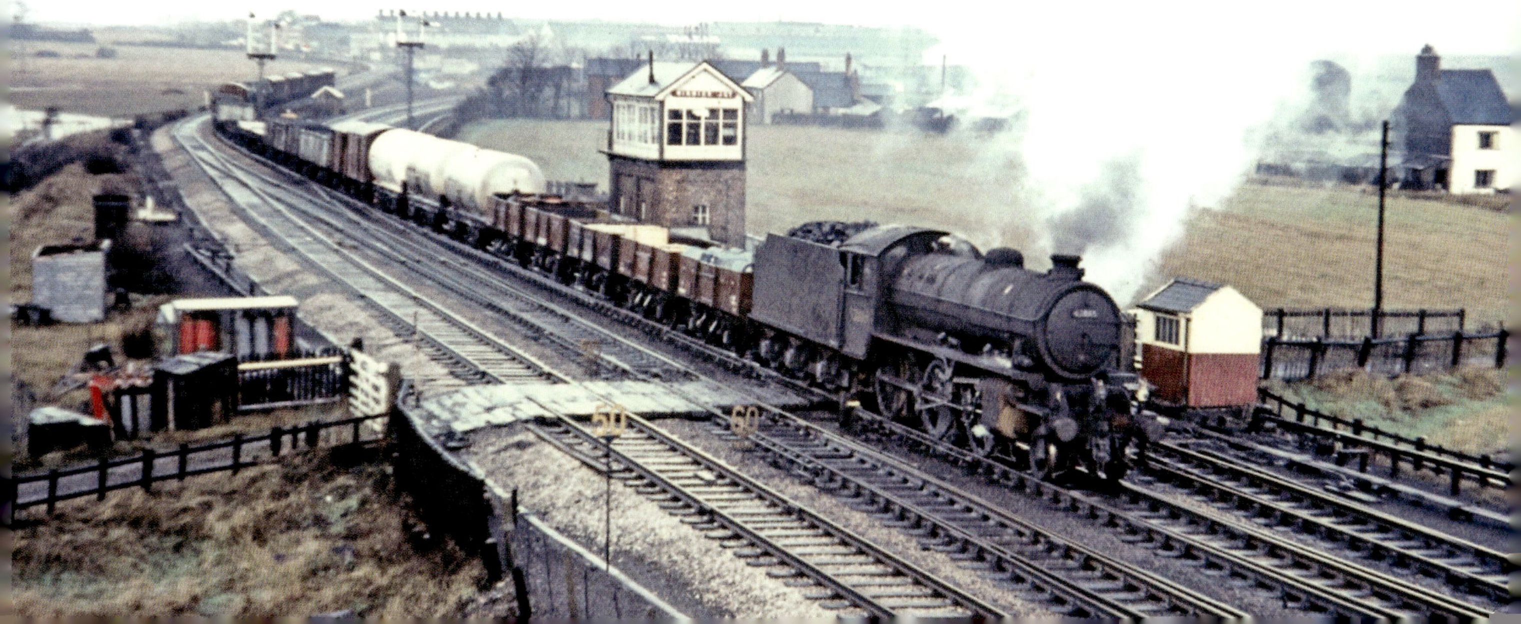

Above. 'Travelling at full tilt' would be an exaggeration, but APT class 370 with unit no. 006 at the driving end passed at speed leaning to the curve at Winwick Jct. It was pictured on 24th August 1984, a month when APT was offering an unscheduled irregular relief service to the paying public from Glasgow to Euston (and return). After APT's calamitous introductory press run on 7/12/1981 managerial as well as technical changes were made to the project so I got to see it again on two further occasions, 7/12/1983 (a test run?) and 29/8/1984. By then the tilting sickness problem had been resolved, and though passengers were not permitted to squeeze past the two motor units they no longer divided the train into equal halves. The picture only shows the front two units and the motor units, but the train's formation was almost certainly 2+2+6 as it was when seen at Moore five days later. Freezing fog at Winwick on 7/12/1983 made it impossible to record the rear of the train, but it certainly had only one motor unit in a 2+1+?5 formation. APT performed perfectly on another press run on 12/12/1984, but that was its final fling. With HST proving to be an unqualified success BR had lost interest in APT, and three years on what remained of the trains were museum pieces. *IMG-1415*

Top right.. On the 4th December 1965 class 8F 2-8-0 no. 48293 was caught northbound labouring up the 1 in 88 by Vulcan Bank signal box towards Earlestown with a train of vans but no rear-end assistance. The vantage point for the photo was the end of the up platform of Vulcan Halt (which had closed six months earlier). The primitive wooden station was just yards from the offices of Vulcan Foundry, but most pedestrians crossing the line would have been heading for the works. VF had a remarkable history from its origins in 1832 to its closure in 2002, a decision swiftly followed by eradication of all trace. Its products had been in demand at home and abroad (and I confess to experiencing a twinge on arriving in Uruguay, 6000 miles away, discovering a 4-4-4T with a Vulcan Foundry works plate [no.3136 of 1915]). But what differentiated VF (later English Electric / VF) from other British loco builders – Beyer Peacock and North British Locomotive spring to mind – was its success in making the transition from steam to diesel and electric. In December 1965 the shops off to the left of the picture would have been busy building a further tranche of a hundred class 20 locos for delivery over the next two years, and the forty-two electro-diesels of class 73/1, while beginning work on the fifty locos of class 50. They were VF's final contribution to BR's traction requirements which had also included the class 55 (Deltics) and class 83 electrics, and most of classes 37, 40, and 86. *HF-S29*

Bottom Right. Earlestown station shares the distinction with only one other – Shipley, W Yorks – of having platforms facing three lines in a triangular shape. The station, formerly known as Newton Junction, was one of the original stopping places for the LMR and thus dates from 1830. In 1831 the WNR began its service to Warrington having connected to the LMR by a curve from West Jct, a crucial link (pre-Mersey bridge at Runcorn) with Liverpool. After absorption by the GJR in 1835 a matching East Jct also opened the way for through running from Manchester. Passenger services have dwindled on the west-to-south route, to the extent that the line is singled and a platform facing no. 3 abandoned. On a bleak early evening on 10th September 1982 class 40 diesel loco no. 40141 is seen here at platform 5 opposite no. 4 with the 17.45 service from Manchester Victoria to Bangor, with the sharpness of the curve very evident. Platforms 1 and 2 serve the LMR line which was electrified in 2015, but when the wires went up on the WCML north of Weaver Jct in the early 1970s it was decided to include also the pre-1864 route to the north through Earlestown, just in case. *IMG-0934-2*

Abbreviations

APT	Advanced Passenger Train	LMS	London, Midland & Scottish
'Black 5'	Stanier class 5MT 4-6-0	LNWR	London & North Western Railway
BJt	Birkenhead Railway (GWR-LNWR Joint) (1860-1923)	LYR	Lancashire & Yorkshire Railway
		MDHB	Mersey Docks and Harbour Board
BLCJR	Birkenhead, Lancashire and Cheshire Joint Railway (1847-59)	Mid.	Midland Railway
		MR	Mersey Railway
BREL	British Rail Engineering Ltd	MSC	Manchester Ship Canal
CBR	Chester & Birkenhead Railway (1840-7)	MS&LR	Manchester, Sheffield & Lincolnshire Railway
CLC	Cheshire Lines Committee		
CME	Chief Mechanical Engineer	MSJAR	Manchester, South Junction and Altrincham Railway
DMU	diesel multiple unit		
ECS	empty coach stock	MT	mixed traffic
EE	English Electric	NWLR	North Wales & Liverpool Railway
EMU	electric multiple unit	p.w.	permanent way (track)
ETH (eth)	electrical train heating	SHCRC	St Helens Canal and Railway Company
GCR	Great Central Railway	SHRGR	St Helens and Runcorn Gap Railway
GJR	Grand Junction Railway	TOPS	(BR's) Total Operations Processing System
GLR	Garston & Liverpool Railway	VF	Vulcan Foundry
GNR	Great Northern Railway	WCML	West Coast Main Line
HST	High Speed Train	WD	War Department
LC&SR	Liverpool, Crosby and Southport Railway	WM&CQR	Wrexham Mold and Connah's Quay Railway
LM	London Midland (region)	WNR	Warrington and Newton Railway
LMR	Liverpool and Manchester Railway	WSR	Warrington and Stockport Railway

Sources and Bibliography

Anderson, P., *An Illustrated History of Liverpool's Railways* (Irwell Press, 1996)

Banks, C. (Comp.), *The Lads From Liverpool* (J.Corkill & P. Hanson) (Silver Link, 2015)

Bolger, P., *Merseyside & District Railway Stations* (Bluecoat, 1994)

British Rail - Main Line, Gradient Profiles (Ian Allan, n.d.)

Brown, J., *Liverpool & Manchester Railway Atlas* (Crécy, 2021)

Earnshaw, A and Aldridge, B., *British Railways Road Vehicles 1948-1968* (Atlantic Transport and Trans Pennine Publishers, 1997)

Edgar, G., *Industrial Locomotives and Railways of the North West of England* (Amberley 2018)

Featherstone, R., and Foster, R., "Chester General", *British Railways Illustrated* March 1994

Hendry, R., Preston and Hendry, R., Powell, *Paddington to the Mersey* (OPC, 1992)

Marsden, C. J. (Ed.), *Locomotive Directory* (Key Publishing, 2021)

Humm, Robert, "The Franco-Crosti Story", *The Railway Magazine* July 2018

Merseyside Railway History Group, *Birkenhead Railways: A Photographic History* (Lightmoor Press, n.d.)

Morris, R. *Railway Mainenance Vehicles and Equipment* (Amberley, 2017)

Nicholls, J., Clough, D., Ratcliffe, D., "Focus on Warrington", *Traction* March/April 2020

Nock, O. S., *The London & North Western Railway* (Ian Allan, 1960)

Norton, P. A., *Railways and Waterways to Warrington* (Cheshire Libraries, 1984)

Pearce, Kenn, *Shed Side on Merseyside* (Sutton, 1997)

Pigott, N., "Water Troughs: Making a Splash", *The Railway Magazine* May 2018

Pigott, N., "Towers of Strength: the Coaling Plant Story", *The Railway Magazine* April-May 2019

Pratt, M., https://www.derbysulzers.com/birkenheadbh.html

Reed, B., *Crewe Locomotive Works and Its Men* (David and Charles, 1982)

Rees, P., "Liverpool's Edge Hill", *The Railway Magazine*, Nov 1977)

Shannon, P., and Hillmer, J., *British Railways Past and Present, No. 39, Liverpool and Wirral* (Past and Present, 2002)

Wellbourn, N., *Lost Lines: Liverpool and the Mersey* (Ian Allan 2008)

https://8dassociation.org/